T0276120

# SpringerBriefs in Molecular Science

## Green Chemistry for Sustainability

**Series editor**

Sanjay K. Sharma, Jaipur, India

More information about this series at http://www.springer.com/series/10045

György Keglevich
Editor

# Milestones in Microwave Chemistry

 Springer

*Editor*
György Keglevich
Budapest University of Technology and
    Economics
Budapest
Hungary

ISSN 2191-5407                    ISSN 2191-5415   (electronic)
SpringerBriefs in Molecular Science
ISSN 2212-9898
SpringerBriefs in Green Chemistry for Sustainability
ISBN 978-3-319-30630-8          ISBN 978-3-319-30632-2   (eBook)
DOI 10.1007/978-3-319-30632-2

Library of Congress Control Number: 2016933216

Printed on acid-free paper

This Springer imprint is published by Springer Nature
The registered company is Springer International Publishing AG Switzerland

# Contents

# Contributors

**Péter Bana** Department of Organic Chemistry and Technology, Budapest University of Technology and Economics, Budapest, Hungary

**Erika Bálint** MTA-BME Research Group for Organic Chemical Technology, Budapest University of Technology and Economics, Budapest, Hungary

**István Greiner** Gedeon Richter Plc, Budapest, Hungary

**György Keglevich** Department of Organic Chemistry and Technology, Budapest University of Technology and Economics, Budapest, Hungary

**Nóra Zs. Kiss** Department of Organic Chemistry and Technology, Budapest University of Technology and Economics, Budapest, Hungary

# Introduction

These days microwave (MW)-assisted chemistry has become an integrant part of environmentally-friendly ("green") chemistry. In a part of the cases, the MW accomplishment offers special advantages compared to the thermal variation. Besides the general benefits of shorter reaction times, higher yields, and selectivity, certain reactions take place only under MW conditions. An additional advantage is that a number of MW-assisted organic chemical transformations may be performed under solvent-free conditions.

In this publication, we tried to provide information that represents hot, but less reviewed topics. After the overview of the developments of the three decades of the MW discipline, fashionable reactions, such as multicomponent reactions, condensations, coupling reactions, and cycloadditions are surveyed. This chapter is followed by a summary of the results attained in the field of organophosphorus chemistry. Theoretical calculations allowed the interpretation of the results and considerations on the scope and limitations of the possible use of the MW technique. Possible simplification of catalysts or catalytic systems under MW conditions are also discussed.

The last part is on the interpretation of advantageous reaction outcomes encountered in MW-assisted organic chemistry on the basis of assumed nonthermal and thermal effects. The different explanations for the rate enhancing effects are discussed critically via well-selected examples. The currently accepted theories are presented in a way to elucidate the adequate phenomena, and the most common misconceptions. Modern experimental techniques in MW chemistry reveal the fundamental role of temperature in interpreting the MW effects. Thermal effects can be differentiated between macroscopic and microscopic effects, both of which are discussed in detail with illustrative examples. The concept of nonthermal MW effects is critically reviewed.

# Chapter 1
# The Spread of the Application of the Microwave Technique in Organic Synthesis

Erika Bálint and György Keglevich

**Abstract** The first chapter summarizes the birth and spread of the application of the microwave (MW) technique in organic syntheses placing the stress on the development of the MW equipment. These days professional batch and continuous flow reactors are available, and the application is knocking at the door of industry.

**Keywords** Microwave · Batch reactors · Continuous reactors

These days, the protection of our environment and our health is becoming increasingly important due to the worldwide spread of green chemistry. According to the 12 principles of green chemistry [1], preparation and development of environmentally-friendly and harmless products and technologies are the main tasks. In this context, the application of the microwave (MW) technique in organic, inorganic, medicinal, analytical and polymer chemistry has spread fast [2–8].

The first domestic microwave oven was introduced by at the end of 1955, but the widespread use of these ovens in households occurred during the 1970s and 1980s. From the middle of 1970s, engineers and researchers started to apply the MW technique in food processing, in the drying industry, in waste remediation and in analytical chemistry. In the latter case, this technique has been used for sample preparation (e.g. digestion, extraction, dissolution, etc.) [9–12]. The first application of microwave irradiation in chemical synthesis was published in 1986 by the groups of Gedye and Giguere [13, 14]. Since then, the number of publications in this field has sharply increased (Fig. 1.1). Most of these publications describe important acceleration of a wide range of organic chemical reactions, excellent repro-

E. Bálint (✉)
MTA-BME Research Group for Organic Chemical Technology,
1521 Budapest, Hungary
e-mail: ebalint@mail.bme.hu

G. Keglevich
Department of Organic Chemistry and Technology, Budapest University
of Technology and Economics, 1521 Budapest, Hungary

© Springer International Publishing Switzerland 2016
G. Keglevich (ed.), *Milestones in Microwave Chemistry*,
SpringerBriefs in Green Chemistry for Sustainability,
DOI 10.1007/978-3-319-30632-2_1

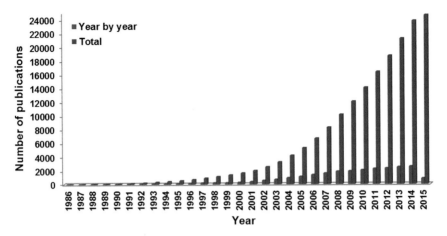

Fig. 1.1 The number of publication on MW-assisted synthesis (1986–2015). Web of Science keyword search on "microwave synthesis"

ducibility, improved yields and less side reactions compared to conventional heating.

Early pioneering experiments were performed in domestic MW ovens, where the irradiation power was controlled generally by on-off cycles of the magnetron, and it was not possible to monitor the inner temperature in a reliable way, thus the reactions were not reproducible. The other problems were on the safety issues of such experiments [15–17]. From the early 2000s, dedicated MW instruments started appearing in market, which are indeed suitable for performing chemical reactions under controlled conditions [2, 3, 18]. All commercially available dedicated MW reactors consist of a MW cavity, magnetic stirrer, sensor probe (IR sensor or fiber optic probe), and software that enables on-line temperature/pressure control by regulating the MW power output.

The MW instruments are classified in two types, monomode (single mode) and multimode MW reactors. The main difference between the two systems is that while in monomode reactors only one reaction vessel can be irradiated, multimode reactors may accommodate several vessels simultaneously.

A monomode instrument has a small compact cavity, where the microwave energy is generated by a single magnetron, and directed through a rectangular waveguide to the reaction mixture, which is positioned at a maximized energy point (Fig. 1.2). A highly homogenous energy field of high power intensity is provided, resulting in exceedingly fast heating rates.

In addition, monomode instruments with a self-tuning circular waveguide are also available (Fig. 1.3). This cavity features multiple entry points for introducing the microwave energy into the vessel.

Multimode reactors have larger cavities, in which the microwaves are reflected from the cavity walls, and distributed in a rather chaotic manner (Fig. 1.4). The reaction vessels are continuously rotated within the cavity, to provide a steady

**Fig. 1.2** The microwave field distribution in a monomode reactor [3]

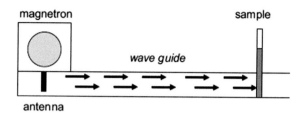

**Fig. 1.3** Circular single-mode cavity [2]

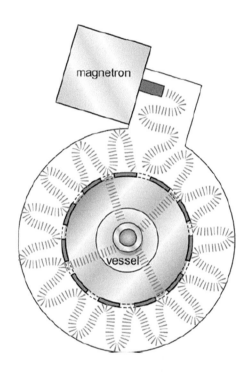

**Fig. 1.4** The microwave field distribution in a multimode parallel synthesis reactor [3]

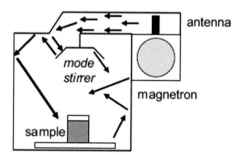

**Fig. 1.5** The microwave field distribution in a multimode single-batch reactor (top view)

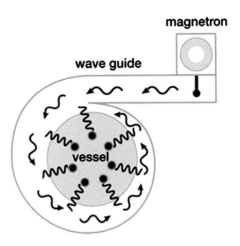

energy distribution. Multimode instruments allow conveniently for parallel syntheses or scale-up. These reactors can host different rotors which are used for parallel reactions in a scale range from several μl up to multi g synthesis in 100 mL reaction vessels.

There is another type of multimode reactor containing a circular waveguide, where various modes of the electromagnetic waves interact with the vessel content at different spots for efficient heating of larger scales (Fig. 1.5). A single few liter vessel is positioned in the cavity, which provides optimal heating rates for large volumes due to the relatively high field density (compared to common multimode microwave oven shown in Fig. 1.4). This kind of multimode reactor is applied for single-batch scale-up procedure, if up to 2 kg of product is required.

Special MW reactors are also known, where the microwave is combined with other techniques, such as UV, ultrasound or high pressure systems (e.g. supercritical reactor) [2].

The scale-up of MW-assisted reactions is of specific interest in many industrial laboratories. The safety limitations of using large batch reactors have promoted the development of continuous flow or stopped-flow MW reactors [19, 20]. These reactors usually comprise three parts, such as the dispensing units for the starting reagents, the MW cavity and the product collector (Fig. 1.6). The reagents are pumped using a HPLC pump or even two pumps. The pressure is controlled by a back-pressure regulator, and the temperature is monitored using a fiber optic sensor or a built-in IR sensor. Usually, the reactors are made from Pyrex or Teflon. The efficiency of the continuous flow MW systems can be increased by using parallel reactors.

Nowadays, there are many types of continuous flow MW reactors, which include a normal flask or tube [21], a fixed bed turbular coil [22–24], an Ω- or U-shaped tube [25–28], a filled column [22, 24, 29] (Fig. 1.7), a spiral glass tube [21, 30–32] (e.g. Emry-type reactor [33] (Fig. 1.8)), a mixed tube [34] (Fig. 1.9) or a capillary reactor [27, 28, 35–37].

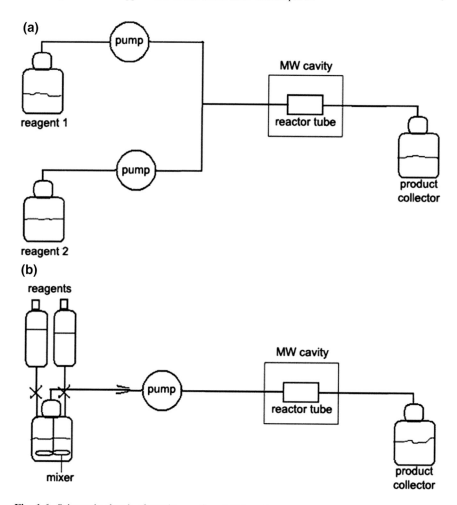

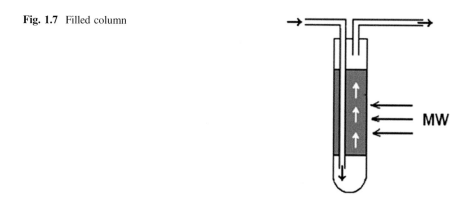

**Fig. 1.6**   Schematic sketch of continuous flow MW reactors

**Fig. 1.7**   Filled column

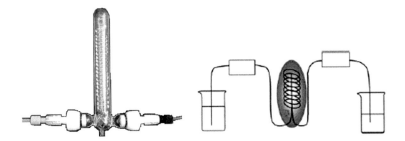

**Fig. 1.8** Emry-type reactor

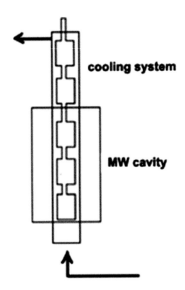

**Fig. 1.9** Mixed tube reactor

There is also a continuous equipment to carry out MW-assisted reaction of solid components (Fig. 1.10) [38, 39].

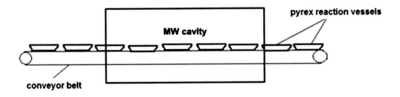

**Fig. 1.10** Continuous microwave reactor for solid-phase reaction

**Fig. 1.11** Isothermal MW
reactor

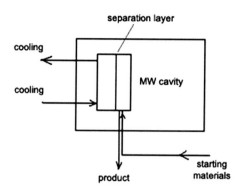

Continuous isothermal MW reactor is also known, which is suitable for implementation of isothermal reactions (Fig. 1.11) [40].

Several MW-assisted continuous flow accomplishments on g or kg scale have been reported in the literature [19, 41–54]. Their capacity may reach 500 kg product per day [55].

## 1.1   Conclusions

In summary, the revolutionary spread of the MW technique resulted in an enormous development in organic chemistry. The appearance of dedicated MW reactors was a "sine qua none" of the new achievements. The mono- and multimode MW batch reactors make possible laboratory scale syntheses, while suitable continuous flow reactors even larger scale production.

## References

1. Anastas PT, Warner JC (1998) Green chemistry: theory and practice. Oxford University Press, New York
2. De La Hoz A, Loupy A (eds) (2012) Microwaves in organic synthesis, vol 1. 3rd edn. Wiley-VCH, Weinheim. doi:10.1002/9783527651313
3. Kappe CO, Stadler A, Dallinger D (2012). In: Mannhold R, Kubinyi H, Folkers G (eds) Microwaves in organic and medicinal chemistry, 2nd edn. Wiley-VCH, Weinheim. doi:10.1002/9783527647828
4. Ameta SC, Punjabi PB, Ameta R, Ameta C (eds) (2014) Microwave-assisted organic synthesis: a green chemical approach. CRC Press, New York
5. Chemat F, Cravotto G (eds) (2013) Microwave-assisted extraction for bioactive compounds: theory and practice. Food engineering series. Springer, New York. doi:10.1007/978-1-4614-4830-3
6. Horikoshi S, Serpone N (eds) (2013) Microwaves in nanoparticle synthesis: fundamentals and applications. Wiley-VCH, Weinheim. doi:10.1002/9783527648122

7. Fang Z, Smith R, Qi X (eds) (2015) Production of biofuels and chemicals with microwave. Biofuels and biorefineries, vol 3. Springer, New York. doi:10.1007/978-94-017-9612-5_1
8. Kempe K, Becer CR, Schubert US (2011) Microwave-assisted polymerizations: recent status and future perspectives. Macromol 44:5825–5842. doi:10.1021/ma2004794
9. Smith FE, Arsenault EA (1996) Microwave-assisted sample preparation in analytical chemistry. Talanta 43:1207–1268. doi:10.1016/0039-9140(96)01882-6
10. Eskilsson SC, Björklund E (2000) Analytical-scale microwave-assisted extraction. J Chromatogr A 902:227–250. doi:10.1016/S0021-9673(00)00921-3
11. Nóbrega JA, Trevizan LC, Araújo GCL, Nogueira ARA (2002) Focused-microwave-assisted strategies for sample preparation. Spectrochim Acta B 57:1855–1876. doi:10.1016/S0584-8547(02)00172-6
12. Chen L, Song D, Tian Y, Ding L, Yu A, Zhang H (2008) Application of on-line microwave sample-preparation techniques. TrAC-Trend Anal Chem 27:151–159. doi:10.1016/j.trac.2008.01.003
13. Gedye R, Smith F, Westaway K, Ali H, Baldisera L, Laberge L, Rousell J (1986) The use of microwave ovens for rapid organic synthesis. Tetrahedron Lett 27:279–282. doi:10.1016/S0040-4039(00)83996-9
14. Giguere RJ, Bray TL, Duncan SM, Majetich G (1986) Application of commercial microwave ovens to organic synthesis. Tetrahedron Lett 27:4945–4948. doi:10.1016/S0040-4039(00)85103-5
15. Abramovitch RA (1991) Applications of microwave-energy in organic-chemistry. A review Org Prep Proced Int 23:683–711. doi:10.1080/00304949109458244
16. Stadler A, Kappe CO (2000) Microwave-mediated Biginelli reactions revisited. On the nature of rate and yield enhancements. J Chem Soc Perkin Trans 2:1363–1368. doi:10.1039/b002697m
17. Vidal T, Petit A, Loupy A, Gedye RN (2000) Re-examination of microwave-induced synthesis of phthalimides. Tetrahedron 56:5473–5478. doi:10.1016/S0040-4020(00)00445-2
18. Rinaldi L, Carnaroglio D, Rotolo L, Cravotto G (2015) A microwave-based chemical factory in the lab: from milligram to multigram preparations. J Chem 2015:1–8. doi:10.1155/2015/879531
19. Moseley JD (2010) Microwave heating as a tool for process chemistry. In: Leadbeater N (ed) Microwave heating as a tool for sustainable chemistry. CRC Press, New York, pp 105–147. doi:10.1002/cssc.201100003
20. Keglevich G, Sallay P, Greiner I (2008) Continuous flow microwave reactors. Hung Chem J 63:278–283
21. Bonaccorsi L, Proverbio E (2008) Influence of process parameters in microwave continuous synthesis of zeolite LTA. Micropor Mesopor Mat 112:481–493. doi:10.1016/j.micromeso.2007.10.028
22. Bo L, Quan X, Chen S, Zhao H, Zhao Y (2006) Degradation of p-nitrophenol in aqueous solution by microwave assisted oxidation process through a granular activated carbon fixed bed. Water Res 40:3061–3068. doi:10.1016/j.watres.2006.06.030
23. Uy SF, Easteal AJ, Farid MM, Keam RB, Conner GT (2005) Seaweed processing using industrial single-mode cavity microwave heating: a preliminary investigation. Carbohyd Res 340:1357–1364. doi:10.1016/j.carres.2005.02.008
24. Bagley MC, Jenkins RL, Lubinu MC, Mason C, Wood R (2005) A simple continuous flow microwave reactor. J Org Chem 70:7003–7006. doi:10.1021/jo0510235
25. Khadilkar BM, Madyar VR (2001) Scaling up of dihydropyridine ester synthesis by using aqueous hydrotrope solutions in a continuous microwave reactor. Org Process Res Dev 5:452–455. doi:10.1021/op010026q
26. Pillai UR, Sahle-Demessie E, Varma RS (2004) Hydrodechlorination of chlorinated benzenes in a continuous microwave reactor. Green Chem 6:295–298. doi:10.1039/b403366c
27. He P, Haswell SJ, Fletcher PDI (2005) Efficiency, monitoring and control of microwave heating within a continuous flow capillary reactor. Sensor Actuat B-Chem 105:516–520. doi:10.1016/j.snb.2004.07.013

28. He P, Haswell SJ, Fletcher PDI (2004) Microwave-assisted Suzuki reactions in a continuous flow capillary reactor. Appl Catal A-Gen 274:111–114. doi:10.1016/j.apcata.2004.05.042
29. Kabza KG, Chapados BR, Getswicki J, McGrath JL (2000) Microwave-induced esterification using heterogeneous acid catalyst in a low dielectric constant medium. J Org Chem 65:1210–1214. doi:10.1021/jo990515c
30. Correa R, Gonzalez G, Dougar V (1998) Emulsion polymerization in a microwave reactor. Polymer 39:1471–1474. doi:10.1016/S0032-3861(97)00413-8
31. Cáceres A, Jaimes M, Chávez G, Bravo B, Ysambertt F, Márquez N (2005) Continuous system with microwave irradiation to obtain alkyl benzoates. Talanta 68:359–364. doi:10.1016/j.talanta.2005.08.067
32. Cablewski T, Faux AF, Strauss CR (1994) Development and application of a continuous microwave reactor for organic synthesis. J Org Chem 59:3408–3412. doi:10.1021/jo00091a033
33. Wilson NS, Sarko CR, Roth GP (2004) Development and applications of a practical continuous flow microwave cell. Org Process Res Dev 8:535–538. doi:10.1021/op034181b
34. Bonnet C, Estel L, Ledoux A, Mazari B, Louis A (2004) Study of the thermal repartition in a microwave reactor: application to the nitrobenzene hydrogenation. Chem Eng Proc 43:1435–1440. doi:10.1016/j.cep.2003.07.003
35. Shore G, Morin S, Organ MG (2006) Catalysis in capillaries by Pd thin films using microwave-assisted continuous-flow organic synthesis (MACOS). Angew Chem Int Ed 45:2761–2766. doi:10.1002/anie.200503600
36. Comer E, Organ MG (2005) A microcapillary system for microwave assisted, high throughput synthesis of molecular libraries. Chem Eur J 11:7223–7227. doi:10.1002/chem.200500820
37. Comer E, Organ MG (2005) A microreactor for microwave-assisted capillary (continuous flow) organic synthesis (MACOS). J Am Chem Soc 127:8160–8167. doi:10.1021/ja0512069
38. Esveld E, Chemat F, van Haveren J (2000) Pilot scale continuous microwave dry-media reactor—Part 1: Design and modeling. Chem Eng Technol 23:279–283. doi:10.1002/(SICI)1521-4125(200003)23:3<279:AID-CEAT279>3.0.CO;2-P
39. Esveld E, Chemat F, Van Haveren J (2000) Scale continuous microwave dry-media reactor—Part II: Application to waxy esters production. Chem Eng Technol 23:429–435. doi:10.1002/(SICI)1521-4125(200005)23:5<429:AID-CEAT429>3.0.CO;2-T
40. Jachuck RJJ, Selvaraj DK, Varma RS (2006) Process intensification: oxidation of benzyl alcohol using a continuous isothermal reactor under microwave irradiation. Green Chem 8:29–33. doi:10.1039/b512732g
41. Singh BK, Kaval N, Tomar S, Eycken EVd, Parmar VS (2008) Transition metal-catalyzed carbon–carbon bond formation Suzuki, Heck, and Sonogashira reactions using microwave and microtechnology. Org Process Res Dev 12:468–474. doi:10.1021/op800047f
42. Baxendale IR, Hayward JJ, Ley SV (2007) Microwave reactions under continuous flow conditions. Comb Chem High Throughput Screen 10:802–836. doi:10.2174/138620707783220374
43. Glasnov TN, Kappe CO (2007) Microwave-assisted synthesis under continuous-flow conditions. Macromol Rapid Commun 28:395–410. doi:10.1002/marc.200600665
44. Ullah F, Samarakoon T, Rolfe A, Kurtz RD, Hanson PR, Organ MG (2010) Scaling out by microwave-assisted, continuous flow organic synthesis (MACOS): Multi-gram synthesis of bromo- and fluoro-benzofused sultams benzthiaoxazepine-1,1-dioxides. Chem Eur J 16:10959–10962. doi:10.1002/chem.201001651
45. Dressen MHCL, van de Kruijs BHP, Meuldijk J, Vekemans JAJM, Hulshof LA (2010) Flow processing of microwave-assisted (heterogeneous) organic reactions. Org Process Res Dev 14:351–361. doi:10.1021/op900257f
46. Bergamelli F, Iannelli M, Marafie JA, Moseley JD (2010) A commercial continuous flow microwave reactor evaluated for scale-up. Org Process Res Dev 14:926–930. doi:10.1021/op100082w

47. Bagley MC, Fusillo V, Jenkins RL, Lubinu MC, Mason C (2010) Continuous flow processing from microreactors to mesoscale: the Bohlmann-Rahtz cyclodehydration reaction. Org Biomol Chem 8:2245–2251. doi:10.1039/B926387J

48. Moseley JD, Lawton SJ (2007) Initial results from a commercial continuous flow microwave reactor for scale-up. Chem Today 25:6–19

49. Benaskar F, Hessel V, Krtschil U, Löb P, Stark A (2009) Intensification of the capillary-based Kolbe-Schmitt synthesis from resorcinol by reactive ionic liquids, microwave heating or a combination thereof. Org Process Res Dev 13:970–982. doi:10.1021/op9000803

50. Dressen MHCL, van de Kruijs BHP, Meduldijk J, Vekemans JAJM, Hulshof LA (2009) From batch to flow processing: racemization of N-acetylamino acids under microwave heating. Org Process Res Dev 13:888–895. doi:10.1021/op9001356

51. Leadbeater NE, Barnard TM, Stencel LM (2008) Batch and continuous-flow preparation of biodiesel derived from butanol and facilitated by microwave heating. Energy Fuels 22:2005–2008. doi:10.1021/ef700748t

52. Smith CJ, Iglesias-Siguenza FJ, Baxendale IR, Ley SV (2007) Flow and batch mode focused microwave synthesis of 5-amino-4-cyanopyrazoles and their further conversion to 4-aminopyrazolopyrimidines. Org Biomol Chem 5:2758–2761. doi:10.1039/b709043a

53. Öhrngren P, Fardost A, Russo F, Schanche JS, Fagrell M, Larhed M (2012) Evaluation of a nonresonant microwave applicator for continuous-flow chemistry applications. Org Process Res Dev 16:1053–1063. doi:10.1021/op300003b

54. Organ MG, Hanson PR, Rolfe A, Samarakoon TB, Ullah F (2011) Accessing stereochemically rich sultams via microwave-assisted, continuous flow organic synthesis (MACOS) scale-out. J Flow Chem 1:32–39. doi:10.1556/jfchem.2011.00008

55. Morschhäuser R, Krull M, Kayser C, Boberski C, Bierbaum R, Püschner PA, Glasnov TN, Kappe CO (2012) Microwave-assisted continuous flow synthesis on industrial scale. Green Process Synth 1:281–290. doi:10.1515/gps-2012-0032

# Chapter 2
# Microwave-Assisted Syntheses in Organic Chemistry

Nóra Zs. Kiss, Erika Bálint and György Keglevich

**Abstract** The second part focuses on the summary of typical organic chemical reactions selected, such as coupling reactions (C–C bond formation reactions, carbon–heteroatom bond formations), condensations (aldol-type-, Claisen-, Knoevenagel reaction), multicomponent reactions (Mannich-, Biginelli-, Hantzsch-, Bucherer–Bergs-, Strecker-, Gewald-, Kabachnik-Fields-, Kindler-, Passerini-, Ugi- and domino reactions), cycloadditions (including Diels–Alder reactions). The authors tried to compile fashionable reactions that have been reviewed less in the past years.

**Keywords** Microwave · Organic chemical reactions · C–C and C–heteroatom coupling reactions · Condensations · Multicomponent reactions · Cycloadditions

## 2.1 Introduction

In the last decades the MW technique has been intensively used to carry out organic reactions of almost all kinds, and has become a useful non-conventional means of performing organic syntheses. This chapter is aimed at giving insights into the new trends of MW-assisted chemistry, placing the stress on the substantial areas of up-to-date synthetic organic chemistry by presenting a selection of the recent literature.

N.Zs.Kiss (✉) · G. Keglevich
Department of Organic Chemistry and Technology, Budapest University of Technology and Economics, 1521 Budapest, Hungary
e-mail: zsnkiss@mail.bme.hu

G. Keglevich
e-mail: gkeglevich@mail.bme.hu

E. Bálint (✉)
MTA-BME Research Group for Organic Chemical Technology, 1521 Budapest, Hungary
e-mail: ebalint@mail.bme.hu

© Springer International Publishing Switzerland 2016
G. Keglevich (ed.), *Milestones in Microwave Chemistry*,
SpringerBriefs in Green Chemistry for Sustainability,
DOI 10.1007/978-3-319-30632-2_2

## 2.2 Coupling Reactions

Carbon-carbon bond forming reactions represent a hot topic in organic chemistry that may benefit from the advantages of MW irradiation resulting in shorter reaction times, as well as simplified accomplishments [1].

Attention has been devoted to develop simple reaction conditions making possible easy product isolations by environmentally benign accomplishments using simple catalysts and green solvents. To face the problems of air sensibility and high cost of typically used *P*-ligands, as well as the difficulties in respect of handling of the reaction mixtures, efforts have been made to develop ligand-free procedures. Driven by environmental concerns, attempts have been made to use water as the solvent.

### 2.2.1 C–C Bond Formation Reactions

#### 2.2.1.1 Heck Reaction

Singh described a versatile phosphine-free protocol for the arylation and benzylation of alkenes under MW irradiation in water (Scheme 2.1) [2]. The Heck reaction was carried out in the presence of Pd(L–proline)$_2$, an air-stable, water-soluble catalyst complex. The substituted olefins (1) were obtained in good yields.

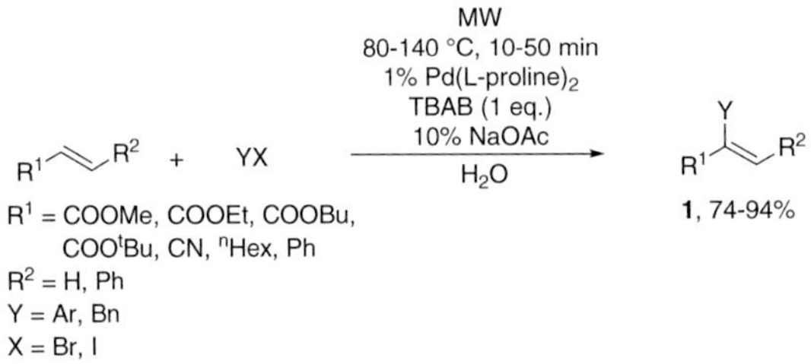

MW
80-140 °C, 10-50 min
1% Pd(L-proline)$_2$
TBAB (1 eq.)
10% NaOAc
H$_2$O

R$^1$ = COOMe, COOEt, COOBu,
COO$^t$Bu, CN, $^n$Hex, Ph
R$^2$ = H, Ph
Y = Ar, Bn
X = Br, I

1, 74-94%

**Scheme 2.1** A phosphine-free Heck reaction

Hervé and Len reported the first MW-assisted, ligand-free cross-coupling reaction of unprotected nucleosides in water. The reaction of 5-iodo-2'-deoxyuridine (2) with various acrylate derivatives was carried out in the presence of Pd(OAc)$_2$ (Scheme 2.2) [3, 4].

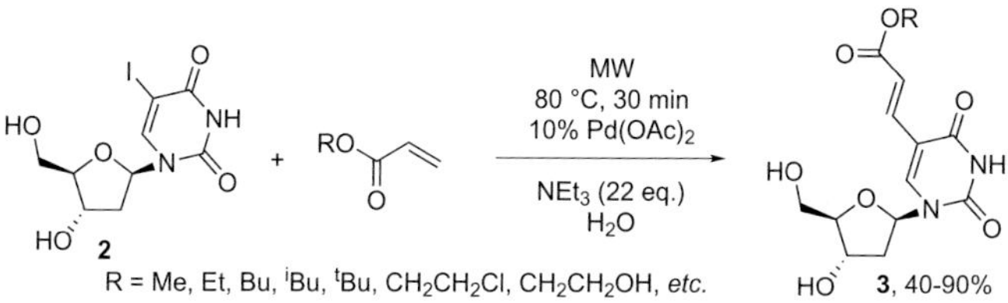

**Scheme 2.2** Ligand-free coupling of a nucleoside in water

The use of task-specific ionic liquids (ILs) is also a "hot topic". A MW-assisted ligand-free and base-free Heck reaction was carried out in a task-specific imidazolium ionic liquid by Dighe and Degani (Scheme 2.3) [5]. The in situ formed palladium complex proved to be an excellent catalyst in terms of activity, selectivity and recyclability under MW irradiation.

**Scheme 2.3** A ligand- and base-free Heck reaction in ionic liquid

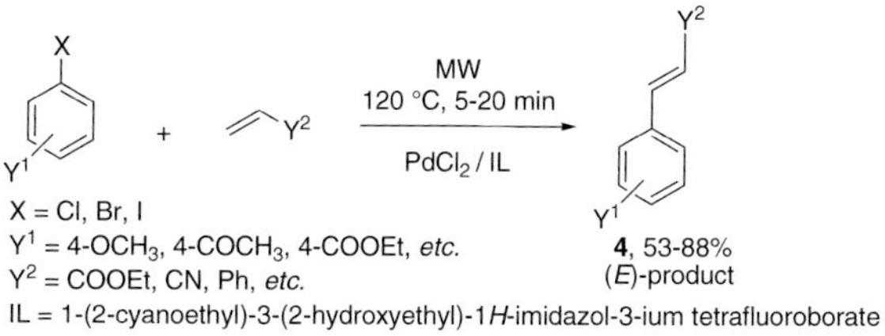

## 2.2.1.2 Suzuki–Miyaura Reaction

A few examples of Suzuki–Miyaura cross-coupling reactions using water as the solvent carried out under MW irradiation can be found in the literature [6–8]. In this series, an up-to-date environmentally friendly synthesis was reported by Cohen and co-workers for the preparation of various 5–substituted thiazoles in the presence of TBAB as a phase transfer catalyst (Scheme 2.4) [9].

**Scheme 2.4** Preparation of 5–substituted thiazoles in water by Suzuki–Miyaura coupling

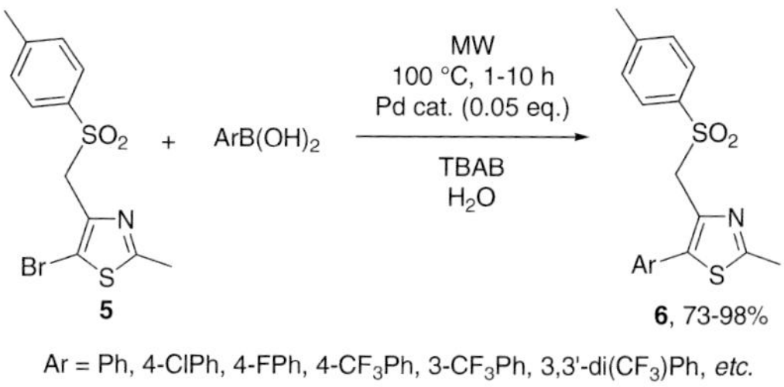

An efficient and solvent-free Suzuki–Miyaura coupling has been developed to form fused tricyclic quinolones using basic alumina as a solid-support and a Pd catalyst under MW irradiation (Scheme 2.5) [10]. The recyclable catalytic system along with the solvent- and base-free conditions, short reaction time and easy handling are remarkable advantages of the synthesis.

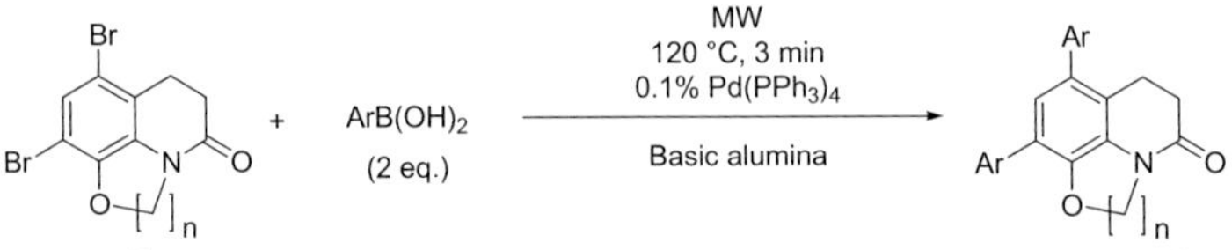

n = 2-4  **7**
Ar = 4-MeOPh, furan-2-yl, thiophen-3-yl, pyridin-3-yl, *etc.*

**Scheme 2.5**  A solvent-free Suzuki–Miyaura coupling

Other examples can also be found, where MW irradiation proved to be beneficial in Suzuki–Miyaura cross-couplings by shortening the reaction times (usually to minutes), and increasing the yields, as compared to those obtained by traditional heating [8, 11–13].

### 2.2.1.3  Hiyama Reaction

A green strategy for the synthesis of biaryls involves a sodium hydroxide activated ligand- and solvent-free Hiyama cross-coupling reaction in the presence of resin-supported Pd nanoparticles under MW heating (Scheme 2.6). A macroporous commercial resin, Amberlite XAD-4, impregnated with Pd nanoparticles (PdNPs) of size 5–10 nm was used efficiently in the coupling of a variety of bromo- and chloroarenes with phenyl-trimethoxysilane. The method of Shah and Kaur benefits from operational simplicity, general applicability and recyclability. The absence of organic solvents, activators and ligands fulfils the requirements of green chemistry [14].

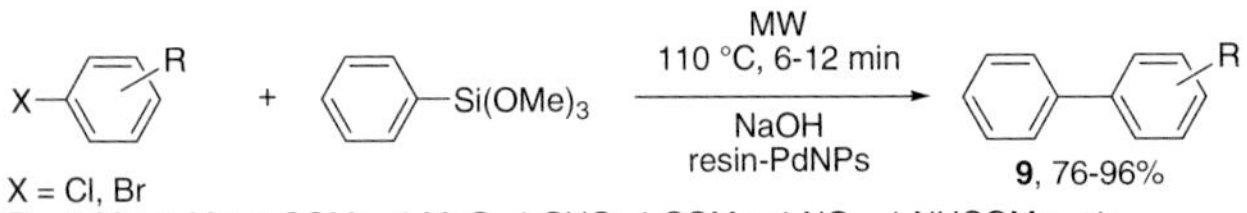

X = Cl, Br
R = 2-Me, 3-Me, 3-COMe, 4-MeO, 4-CHO, 4-COMe, 4-NO$_2$, 4-NHCOMe, *etc.*

**Scheme 2.6**  Ligand- and solvent-free Hiyama cross-coupling to form biaryls

## 2.2.2  Carbon–Heteroatom Bond Formations

Carbon–heteroatom bond formations were also studied intensively under MW-assisted conditions to reduce reaction times, simplify catalyst systems, or eliminate organic solvents.

### 2.2.2.1  Microwave-Assisted C–N Bond Formation

Gupta and Singh described a simple and environmentally-friendly C–N coupling of a wide range of aryl halides and amines under ligand-free and solvent-free MW conditions (Scheme 2.7) [15]. Not only short reaction times were required, but the heterogeneous catalyst applied could be recovered by simple filtration, and could be re-used.

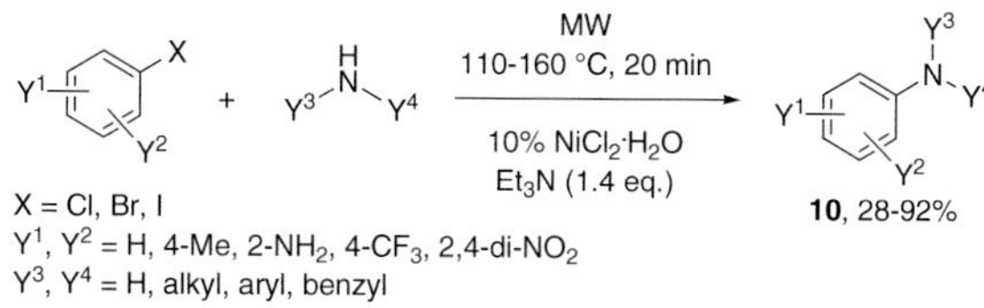

Scheme 2.7  Ligand- and solvent-free C–N coupling of aryl halides and amines

Aryl halides and amines were also subjected to iron/copper co-catalyzed ligand-free reactions under MW irradiation (Scheme 2.8) [16]. It is worth mentioning that the simple reaction conditions were associated with a broad substrate scope.

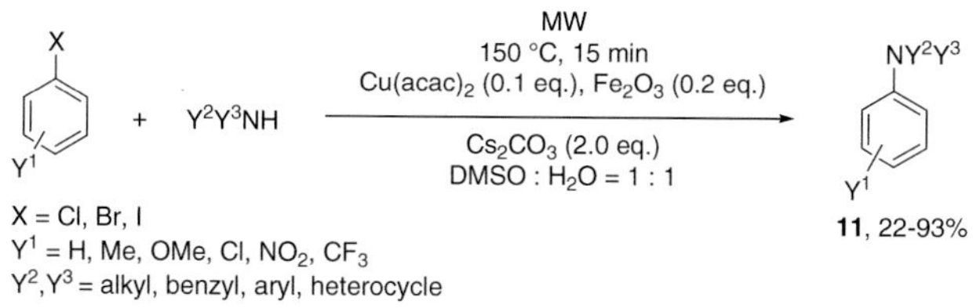

Scheme 2.8  Iron/copper co-catalyzed ligand-free C–N bond formation

Halopyridines and various nitrogen nucleophiles were subjected to a MW-assisted copper-catalyzed cross-coupling without the use of any ligands or solvents (Scheme 2.9) [17].

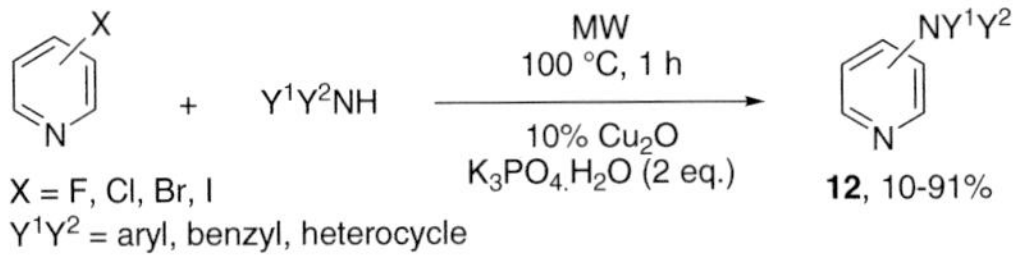

**Scheme 2.9** Ligand- and solvent-free C–N bond formation of pyridine-derivatives

*N*-arylimidazoles of pharmaceutical interest were also prepared by a MW-assisted solvent-free *N*-arylation [18].

### 2.2.2.2  Microwave-Assisted C–P Bond Formation

The Hirao reaction [19] is an important tool for the formation of P–C bond. See also Sect. 3.5. Many publications highlight the beneficial effect of MW irradiation in the Hirao reaction [20–23]. Keglevich and Jablonkai developed the first *P*-ligand- and solvent-free Pd-calayzed coupling of different >P(O)H species with aryl-bromides in the presence of Pd(OAc)$_2$ under MW conditions (Scheme 2.10) [24]. This accomplishment is the first example for *P*-ligand-free Hirao reactions.

**Scheme 2.10**  A novel *P*-ligand-free Hirao reaction

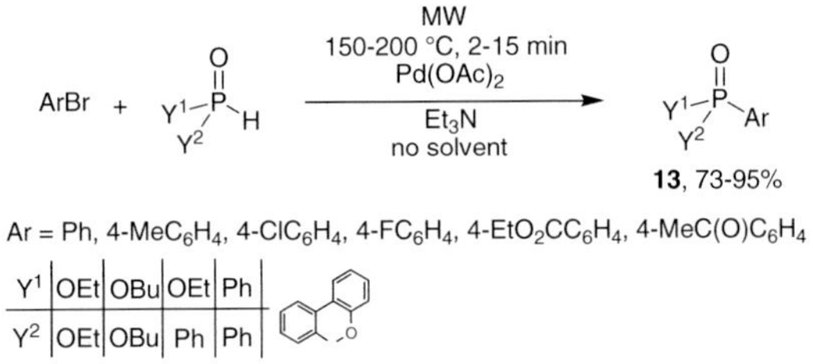

Arylphosphonates, phosphinates or phosphine oxides could all be formed in the coupling reaction of >P(O)H species and aryl halides in the presence of Cu or Ni salts [25]. Starting from the salts of the >P(O)H species, there was no need for any catalysts [26].

## 2.3  Condensations and Multicomponent Reactions

During condensations, two or more molecules are combined, usually in the presence of a catalyst to form the product with the elimination of water or another simple molecule.

Multicomponent reactions are convergent reactions, in which three or more compounds react to form a product, where the majority of the atoms of the components is incorporated in the newly formed product. Most of the classical multicomponent reactions involve the participation of carbonyl compounds and/or their derivatives.

In general, traditional conductive heating methods are used to realize condensations and multicomponent reactions. These methods are often slow, and the conventional heating is not really suitable from the point of view of energy efficiency. The use of MW irradiation is more efficient and ecofriendly to carry out these reactions, as shorter reaction times, enhanced reaction rates, and higher yields can be attained in comparison with conventional heating [27].

In this subchapter, several MW-assisted condensations, as well as multicomponent reactions, such as aldol-, Claisen- and Knoevenagel condensations, Mannich-, Biginelli-, Bucherer-Bergs-, Strecker-, Gewald-, Hantzsch-, Kabachnik-Fields-, Kindler-, Passerini-, Ugi- and domino reactions will be discussed.

## 2.3.1 Aldol-Type Condensations

Aldol condensation is a typical way to form a carbon–carbon bond. In the condensation, an enolizable aldehyde or ketone reacts with a carbonyl compound to form a β–hydroxyaldehyde or β-hydroxyketone, followed by a dehydration step to give a conjugated enone.

A MW-assisted method was developed by Marijani et al. for the synthesis of hydroxy-cyclopentenones (**14**) by the condensation of benzil with ketones carried out in the presence of KOH/EtOH at 180 °C for 2–8 min (Scheme 2.11) [28].

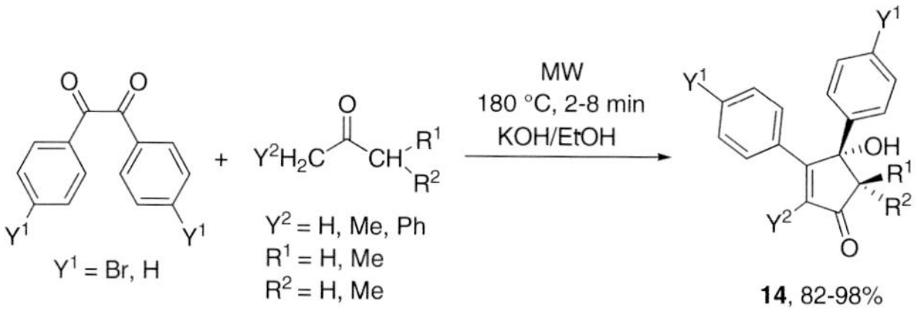

**Scheme 2.11** MW-assisted synthesis of hydroxy-cyclopentenones

The MW-assisted aldol-type condensations of 3-methyl-2-cyclohexenones and aromatic aldehydes were studied using $BiCl_3$ as the catalyst in the absence of any solvent (Scheme 2.12) [29].

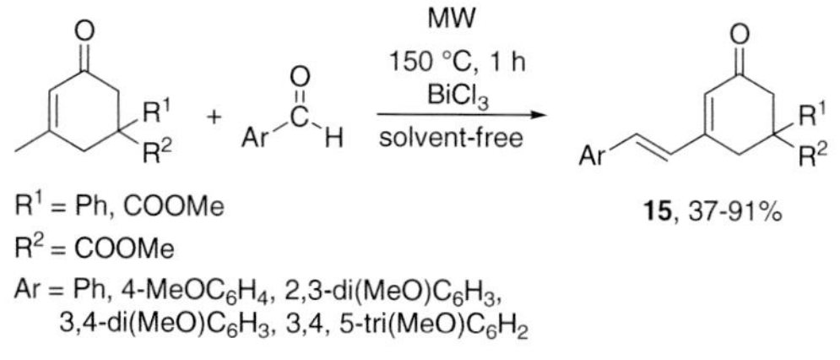

**Scheme 2.12** Condensation of 3-methyl-2-cyclohexenones and aromatic aldehydes

## 2.3.2 Claisen Condensations

The Claisen condensation [30] is the "ester analogue" of the aldol condensation. During the reaction, two esters, or one ester and another carbonyl compound react with each other in the presence of a strong base to form a β-keto ester or a β-diketone.

An ultraviolet absorbent, 4-*tert*-butyl-4′-methoxydibenzoylmethane (trade name Avobenzone) (16) was synthesized by the Claisen condensation of 4-methoxyacetophenone and methyl 4-*tert*-butylbenzoate in a household MW oven using sodium amide as the base, and toluene as the solvent (Scheme 2.13) [31].

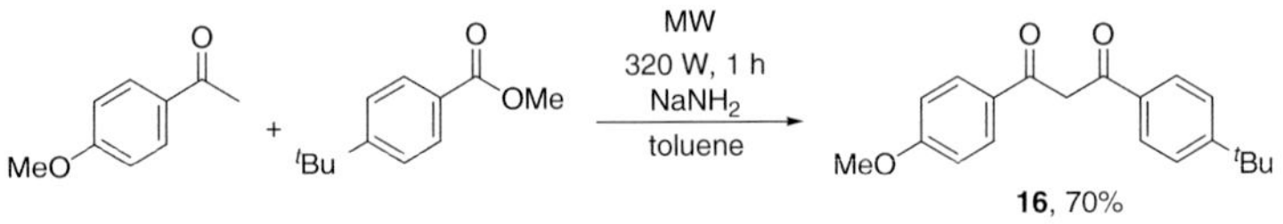

Scheme 2.13 Claisen condensation of 4-methoxyacetophenone and methyl 4-*tert*-butylbenzoate

## 2.3.3 Knoevenagel Condensations

The Knoevenagel reaction [32] is a modified aldol condensation between an aldehyde or ketone, and an active methylene group containing compound in the presence of a base catalyst. The reaction is usually followed by a spontaneous dehydration step resulting in an unsaturated product.

A high nitrogen containing mesoporous carbon nitride (MCN) was applied as a metal-free base catalyst in the Knoevenagel condensation of aromatic aldehydes with ethyl cyanoacetate (Scheme 2.14) [33]. The reactions were performed in toluene under MW irradiation, and the products (17) were obtained in yields of 75–95 %.

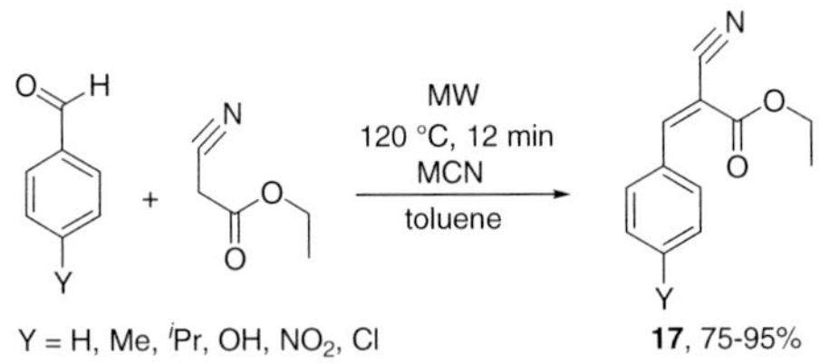

Y = H, Me, *i*Pr, OH, NO$_2$, Cl          17, 75-95%

Scheme 2.14 Knoevenagel condensation of aromatic aldehydes with ethyl cyanoacetate

The condensation of 3-α-carboxy ethylrhodanine (18) with substituted aromatic aldehydes in the presence of sodium acetate in glacial acetic acid was studied under MW irradiation at 150 °C for 10–15 min (Scheme 2.15) [34]. The reactions afforded 5–benzylidene-3-α-carboxy ethylrhodanine derivatives (19) in high yields.

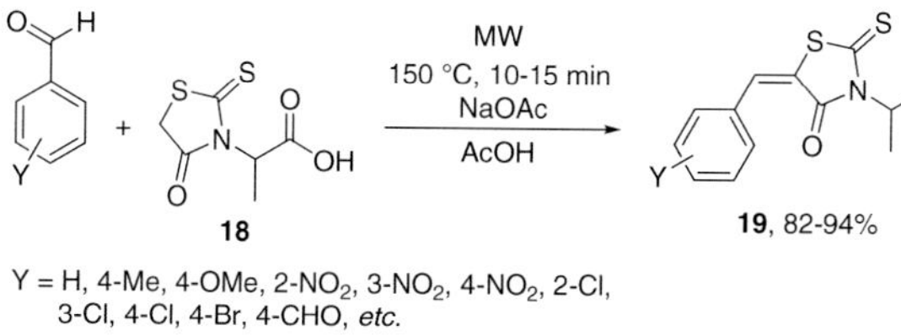

Y = H, 4-Me, 4-OMe, 2-NO$_2$, 3-NO$_2$, 4-NO$_2$, 2-Cl, 3-Cl, 4-Cl, 4-Br, 4-CHO, *etc.*

**Scheme 2.15** Condensation of 3-α-carboxy ethylrhodanine with aromatic aldehydes

The MW-assisted Knoevenagel reactions of 2,5-disubstituted indole-3-carboxaldehydes (**20**) and active methylene group containing compounds were studied by Biradar and Sasidhar (Scheme 2.16) [35]. The reactions were carried out in a household MW oven, in the presence of ammonium acetate under solvent-free conditions. It was found that without catalyst, the yields were very low and sometimes no reaction occurred.

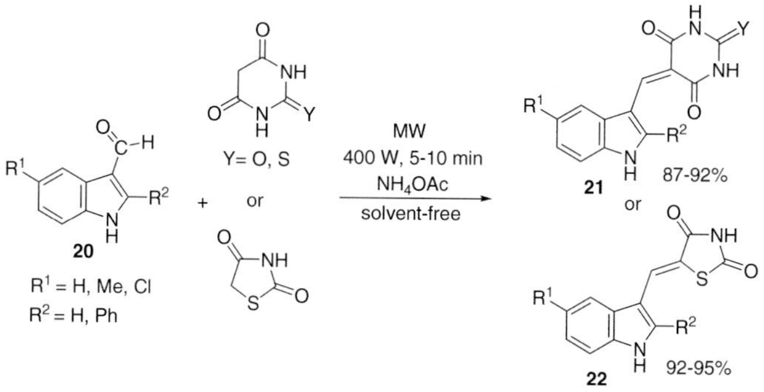

**Scheme 2.16** Knoevenagel reaction of 2,5-disubstituted indole-3-carboxaldehydes

## 2.3.4 *Mannich Reactions*

The Mannich reaction [36] is a three-component condensation, where a primary or secondary amine (or ammonia) reacts with an aldehyde and a ketone. The final product is a β-amino-carbonyl compound, also known as a Mannich base.

β-Amino-carbonyl derivatives (**23**) were synthesized in the three-component condensation of aniline derivatives, aromatic aldehydes and cyclohexanone using CeCl$_3$ as the catalyst under solvent-free and MW conditions (Scheme 2.17) [37].

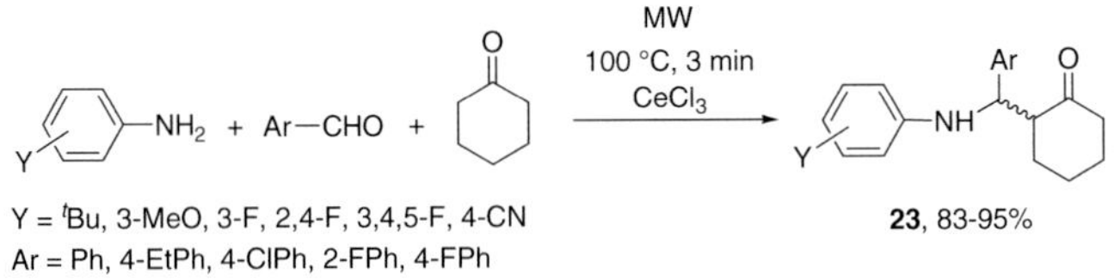

**Scheme 2.17** MW-assisted solvent-free condensation of anilines, aromatic aldehydes and cyclohexanone

MW-assisted Mannich reactions of secondary amine hydrochlorides, paraformaldehyde and substituted acetophenones were studied by Luthman and co–workers (Scheme 2.18) [38]. The reactions were carried out in dioxane on a small (2 mmol) and also on a larger (40 mmol) scale.

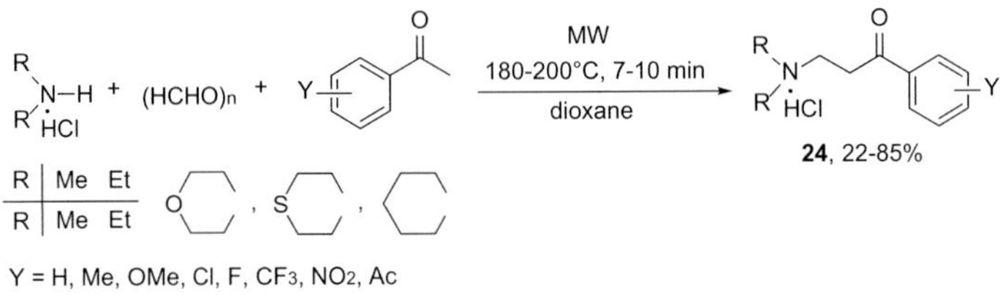

**Scheme 2.18** Mannich reaction of amine hydrochlorides, paraformaldehyde and acetophenones

Mannich-type reactions of secondary amines, aldehydes and acetylene derivatives were investigated by Leadbeater et al. (Scheme 2.19) [39]. The condensations were performed in dioxane, in the presence of CuCl and a small amount of ionic liquid (IL) under MW irradiation. Using IL as the solvent instead of dioxane, a decomposition was observed.

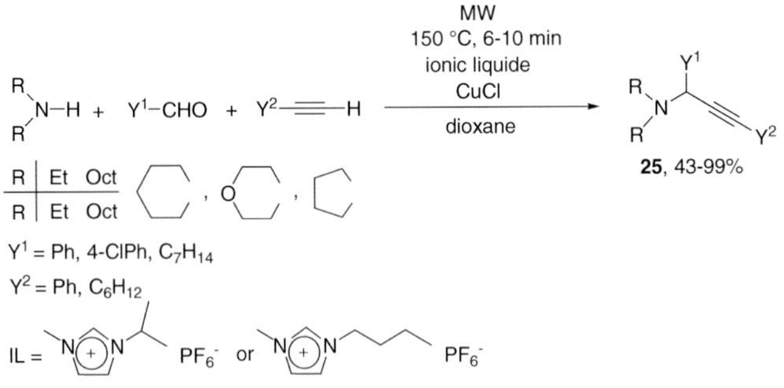

**Scheme 2.19** Mannich-type reaction of secondary amines, aldehydes and acetylenes

The MW-assisted condensation of a 2-hydroxy-chalcone (**26**) was studied with secondary amines and paraformaldehyde (Scheme 2.20) [40]. The reactions were performed in dioxane without any catalyst at 100 °C for 10–45 min, and the corresponding products (**27**) were obtained in yields of 81–97 %.

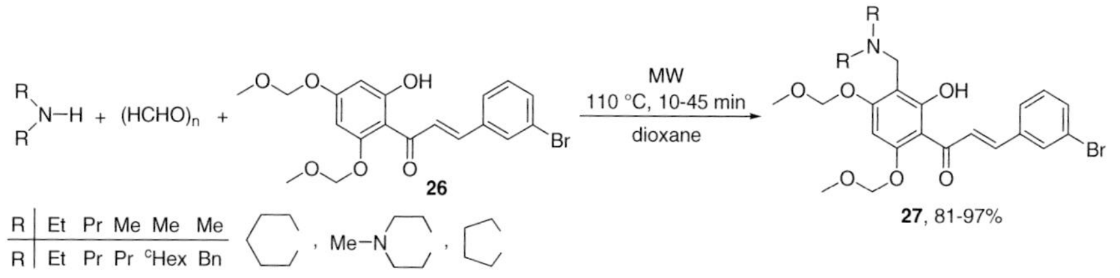

**Scheme 2.20**  Condensation of a 2-hydroxy-chalcone

## 2.3.5  Biginelli Reactions

The Biginelli reaction [41] is a multicomponent one-pot condensation of an aldehyde, a β–keto ester and an urea derivative to afford dihydropyrimidinones, which are of a wide range of pharmaceutical and therapeutic properties [42, 43].

MW-assisted Biginelli reactions of aromatic aldehydes, 1,3-dicarbonyl compounds and urea or thiourea were studied by Japanese researchers (Scheme 2.21) [44]. The condensations were carried out using tributyl borate as the catalyst under solvent-free conditions, and the corresponding dihydropyrimidinones (**28**) were obtained in high yields.

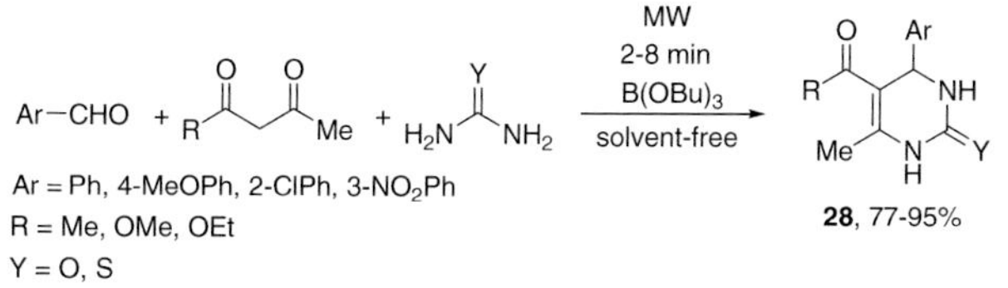

**Scheme 2.21**  Biginelli reaction of aromatic aldehydes, 1,3-dicarbonyl compounds and ureas

Chinese researchers elaborated a fast and solvent-free MW-assisted method for the synthesis of dihydropyrimidinone derivatives (**29**), but in this case, a heteropolyanion-based IL was applied as the catalyst (Scheme 2.22) [45].

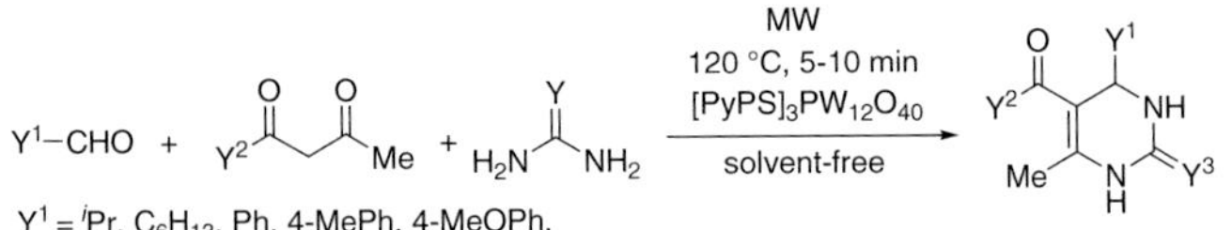


$Y^1$ = $^i$Pr, $C_6H_{13}$, Ph, 4-MePh, 4-MeOPh, 4-ClPh, 4-$NO_2$Ph, 4-HOPh
$Y^2$ = OEt, OMe
$Y^3$ = O, S

**29**, 83-96%

**Scheme 2.22** MW-assisted synthesis of dihydropyrimidinones

There is a good example, where the multicomponent reaction of aromatic aldehydes, acetoacetamine derivatives and ureas was performed under solvent- and catalyst-free conditions (Scheme 2.23) [46]. MW irradiation at 120 °C for 12–16 min furnished the dihydropyrimidinones (**30**) in 70–75 % yields.

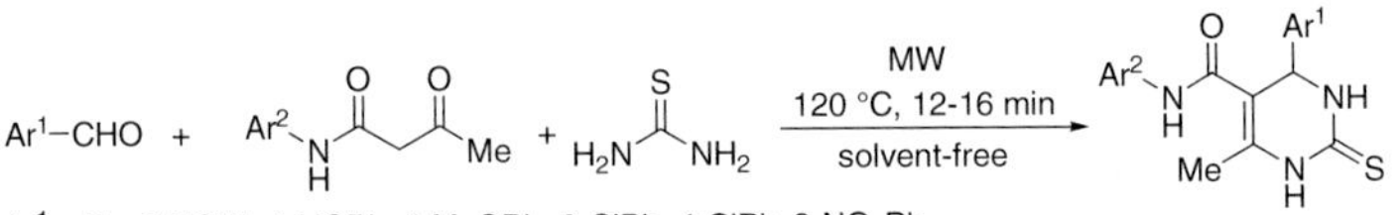


$Ar^1$ = Ph, 2-HOPh, 4-HOPh, 4-MeOPh, 2-ClPh, 4-ClPh, 3-$NO_2$Ph
$Ar^2$ = 4-$ClC_6H_4$, 4-$FC_6H_4$, 1-naphthyl

**30**, 70-75%

**Scheme 2.23** A solvent- and catalyst-free Biginelli reaction under MW irradiation

Fang and Lam reported a modified MW-assisted Biginelli reaction of aromatic aldehydes, 2–oxosuccinic acid and substituted ureas, which led to aryl-oxo-tetrahydropyrimidinyl-carboxylic acid derivatives (**31**) by cyclization accompanied by decarboxylation (Scheme 2.24) [47]. The reactions were performed in THF, and were catalyzed by trifluoroacetic acid (TFA).

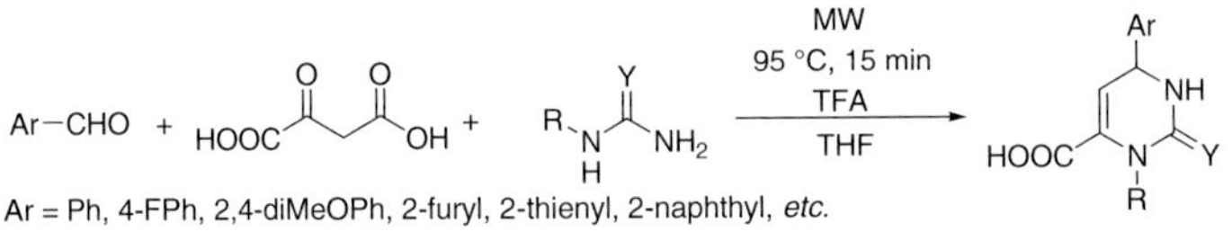


Ar = Ph, 4-FPh, 2,4-diMeOPh, 2-furyl, 2-thienyl, 2-naphthyl, *etc.*
R = H, Me, allyl
Y = O, S

**31**, 71-93%

**Scheme 2.24** Condensation of aromatic aldehydes, 2–oxosuccinic acid and substituted ureas

The synthesis of 3,4-dihydropyrimidin-2(1*H*)-ones (**33**) was studied starting from an IL supported aldehyde (**32**), a β–ketoester and an urea (Scheme 2.25) [48]. HCl was used as catalyst, and the reactions were carried out in the absence of solvent under MW irradiation. The corresponding products (**33**) were obtained in good yields after the cleavage of the IL moiety realized by transesterification with NaOMe/MeOH at reflux.

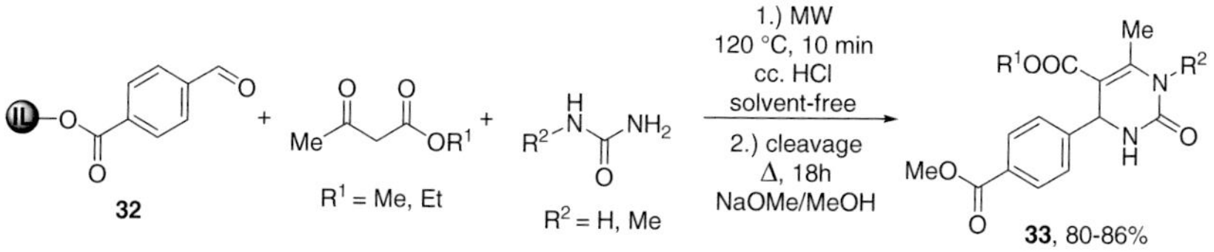

**Scheme 2.25**  Biginelli reaction of IL supported aldehyde, β–ketoesters and ureas

## 2.3.6  Hantzsch Reactions

The Hantzsch dihydropyridine synthesis [49] is a four-component reaction with the participation of an aldehyde, two equivalents of a β-ketoester and a "nitrogen donor", such as ammonium acetate, or ammonia. Subsequent oxidation (or dehydrogenation) may lead to pyridine-3,5-dicarboxylates, which may undergo decarboxylation to yield the corresponding pyridines.

Westman and Öhberg developed a MW-assisted Hantzsch reaction of different aldehydes, β-ketoesters and aqueous ammonium hydroxide (Scheme 2.26) [50]. $NH_4OH$ was used as the reagent, and also as the solvent. After an irradiation at 140–150 °C for 10–15 min, the corresponding dihydropyridines (**34**) were formed in moderate to good yields.

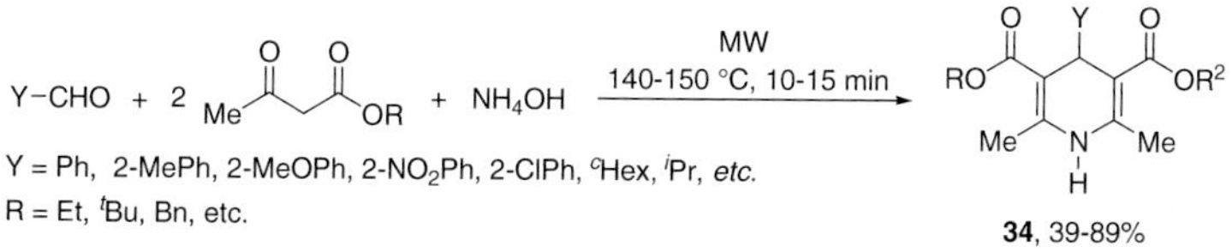

**Scheme 2.26**  Hantzsch reaction of aldehydes, β-ketoesters and aqueous ammonium hydroxide

A bismuth nitrate-catalyzed cyclocondensation was reported by American researchers (Scheme 2.27) [51]. A series of dihydropyridines (**35**) were synthesized using a series of aldehydes, 1,3-diketo compounds and ammonium acetate or amines under solvent-free MW conditions.

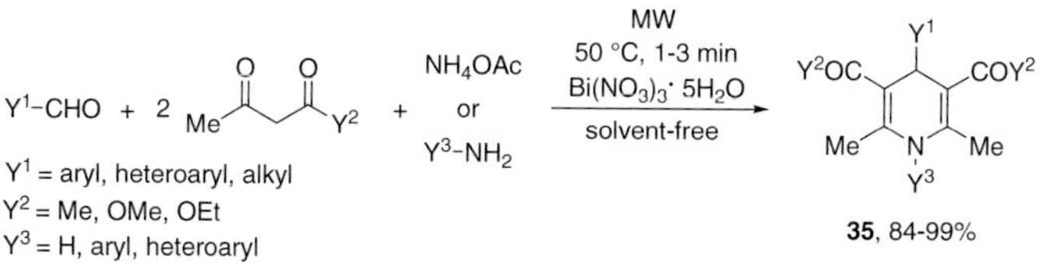

**Scheme 2.27**  A bismuth nitrate-catalyzed cyclocondensation

Silicotungstic acid nanoparticles dispersed in the micropores of Cr-pillared clay (STA/Cr–P) were used as heterogeneous catalysts for the solvent-free synthesis of 1,4–dihydropyridines (**36** or **37**) (Scheme 2.28) [52]. During these reactions, aryl aldehydes or chalcones were reacted with ethyl acetoacetate and ammonium acetate under continuous MW irradiation at 900 W. After regeneration, the STA/Cr-P catalyst was re-usable for several times.

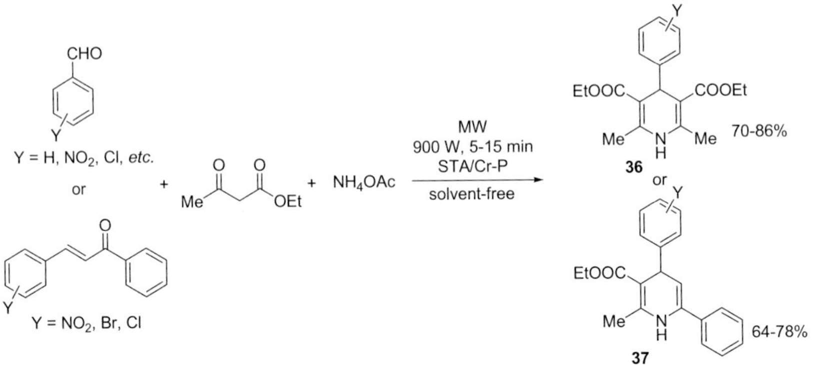

**Scheme 2.28** Synthesis of 1,4-dihydropyridines in the presence of STA/Cr-P catalyst

A MW-assisted synthesis of 1,4-dihydropyridines (**38**) using task-specific ILs as a soluble support was described by Bazureau and co-workers (Scheme 2.29) [48]. In the first step, the functionalized IL phase-bound aldehyde (**32**) was reacted with the β–ketoester and aminocrotonate under solvent-free and MW-assisted conditions. 5-$N$-(2-Hydroxyethyl)pyridinium hexafluoroborate ([PEG$_1$py][PF$_6$]) was used as the IL. Then, the IL support was cleaved from the product by transesterification with NaOMe/MeOH at reflux. The desired compounds (**38**) were obtained in yields of 85–86 %.

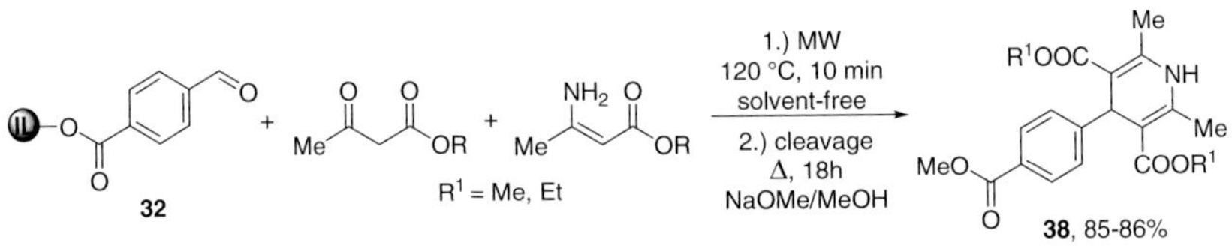

**Scheme 2.29** Hantzsch reaction of IL phase-bound aldehyde, β–ketoester and aminocrotonate

## 2.3.7  Bucherer-Bergs Reactions

The Bucherer-Bergs reaction [53, 54] is a multi-component transformation with the participation of carbonyl compounds (aldehydes or ketones), cyanohydrines or potassium cyanide and ammonium carbonate, which leads to the formation of hydantoins.

5,5-Disubstituted hydantoins (**39**) were obtained in high yields by the condensation of carbonyl derivatives, potassium cyanide and ammonium carbonate in the presence of EtOH/H$_2$O under MW conditions (Scheme 2.30) [55].

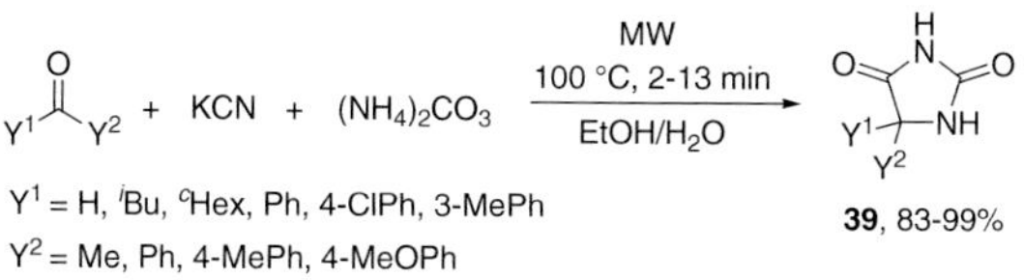

$$Y^1 = H, {}^tBu, {}^cHex, Ph, 4\text{-ClPh}, 3\text{-MePh}$$
$$Y^2 = Me, Ph, 4\text{-MePh}, 4\text{-MeOPh}$$

**39**, 83-99%

**Scheme 2.30**  MW-assisted synthesis of 5,5-disubstituted hydantoins

The synthesis of phenylpiperazine hydantoin derivatives was studied by Polish researchers [56]. The compounds were obtained in four steps, where the first step was the Bucherer-Bergs reaction of acetophenone with potassium cyanide and ammonium carbonate under MW conditions (Scheme 2.31).

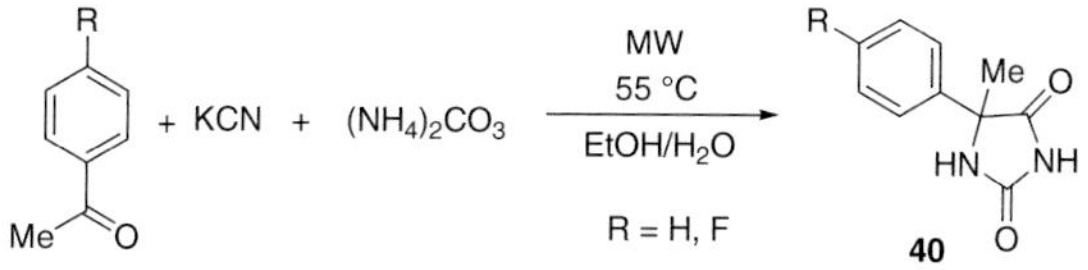

R = H, F

**40**

**Scheme 2.31**  Bucherer-Bergs reaction of acetophenone, potassium cyanide and ammonium carbonate

## 2.3.8  Strecker Reactions

The Strecker synthesis [57] provides an amino acid from an aldehyde or ketone. The oxo component is condensed with ammonium chloride in the presence of potassium cyanide to furnish an α-aminonitrile, which is subsequently hydrolyzed to give the desired amino acid.

The Nafion-Fe-catalyzed Strecker reaction of various aldehydes or ketones with amines and trimethylsilyl cyanide were investigated, and the corresponding α-aminonitriles (**41**) were obtained in yields of 49–97 % under solvent-free MW conditions (Scheme 2.32) [58].

**Scheme 2.32** A Nafion-Fe-catalyzed Strecker reaction

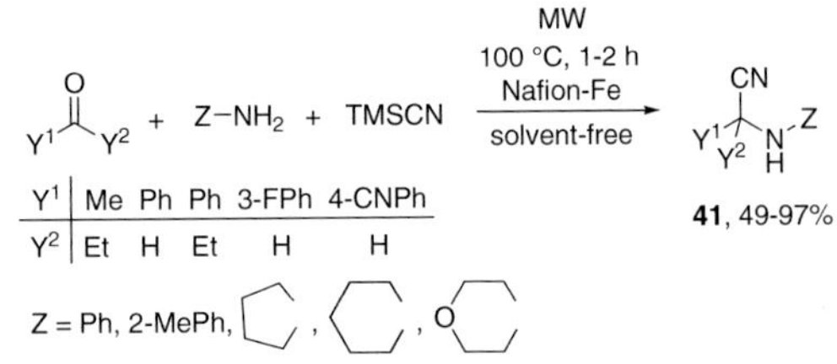
MW
100 °C, 1-2 h
Nafion-Fe
solvent-free
+ Z−NH2 + TMSCN
Y1 Me Ph Ph 3-FPh 4-CNPh
Y2 Et H Et H H
Z = Ph, 2-MePh,
41, 49-97%

A series of α-aminonitriles (**42**) were synthesized via a catalytic Strecker-type reaction of aldehydes, amines and trimethylsilyl cyanide (Scheme 2.33) [59]. The reactions were carried out at low temperature in the presence of Co(II) complex supported on mesoporous SBA-15 under solvent-free MW-assisted conditions.

**Scheme 2.33** A catalytic Strecker-type reaction

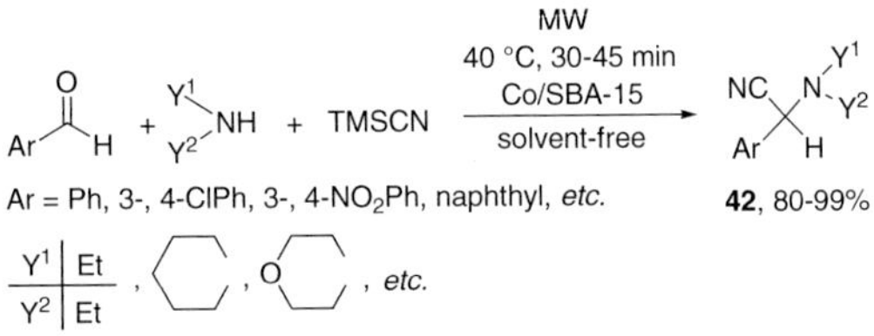
MW
40 °C, 30-45 min
Co/SBA-15
solvent-free
+ TMSCN
Ar = Ph, 3-, 4-ClPh, 3-, 4-NO2Ph, naphthyl, etc.
Y1 Et
Y2 Et
etc.
42, 80-99%

A somewhat Strecker analogous reaction accompanied by decarboxylation was studied by Seidel and co–workers. Proline was reacted with different aldehydes and trimethylsilyl cyanide in butanol under MW irradiation as shown in Scheme 2.34 [60].

**Scheme 2.34** A Strecker analogous reaction

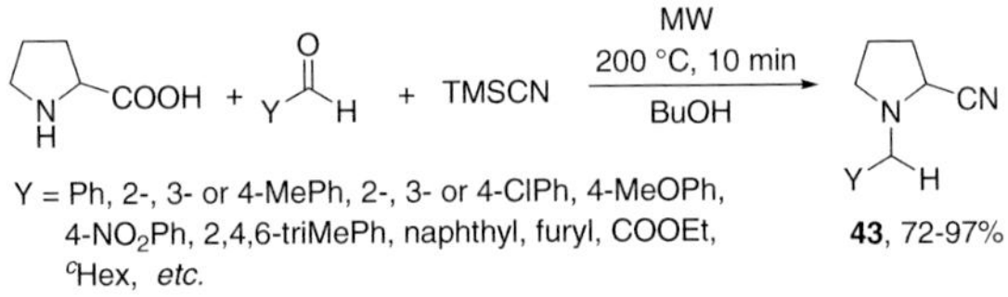
COOH
+ TMSCN
MW
200 °C, 10 min
BuOH
CN
Y = Ph, 2-, 3- or 4-MePh, 2-, 3- or 4-ClPh, 4-MeOPh,
4-NO2Ph, 2,4,6-triMePh, naphthyl, furyl, COOEt,
cHex, etc.
43, 72-97%

## 2.3.9 Gewald Reactions

The Gewald reaction [61] involves the synthesis of 2-aminothiophene derivatives via the multi-component condensation of an α-methylene carbonyl compound, an α-cyanoester and elemental sulfur in the absence of a base.

Kirsh and co-workers developed a MW-assisted procedure for the Gewald reaction of aldehydes, activated nitriles and sulfur (Scheme 2.35) [62]. The condensations were carried out at 70 °C for 20 min using morpholine as the base, and ethanol as the solvent.

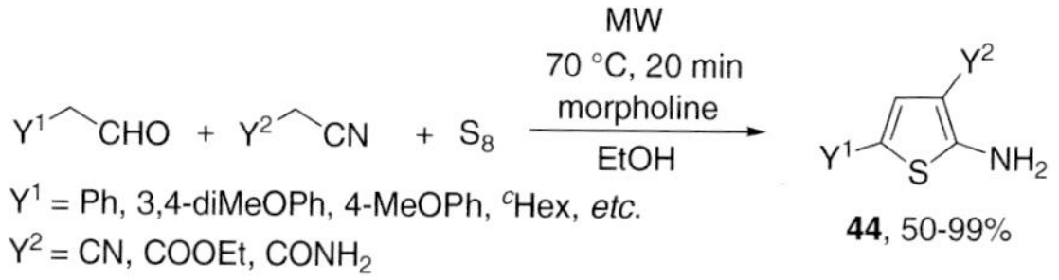

**Scheme 2.35**  MW-assisted Gewald reaction of aldehydes, activated nitriles and sulfur

The condensation of ketones with cyanoacetate or malononitrile and sulfur was studied under MW conditions (Scheme 2.36) [63]. The multicomponent reactions were performed using KF-alumina as the catalyst instead of an organic base, and the 2–aminothiophenes (**45**) were obtained in short times and in yields of 55–92 %.

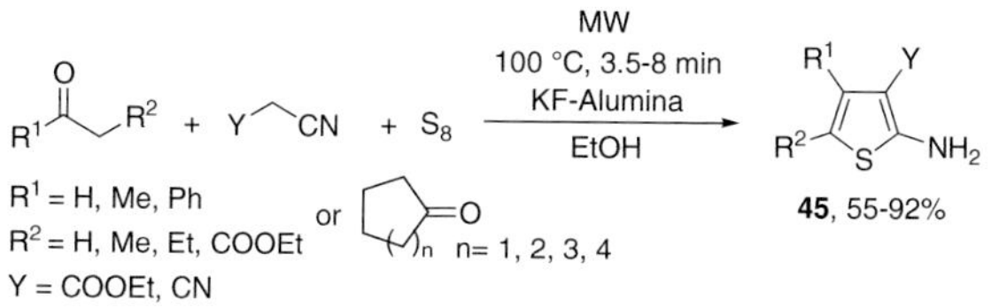

**Scheme 2.36**  KF-alumina catalyzed Gewald reaction

A guanidine-catalyzed Gewald condensation was reported (Scheme 2.37) [64]. In the course of the reaction, a mixture of cyclopentanone, 2-cyano-*N*-*o*-tolylacetamide and elemental sulfur was irradiated continuously in a MW reactor in the presence of a 1,1,3,3-tetramethylguanidine lactate IL. This reaction was also carried out in ethanol.

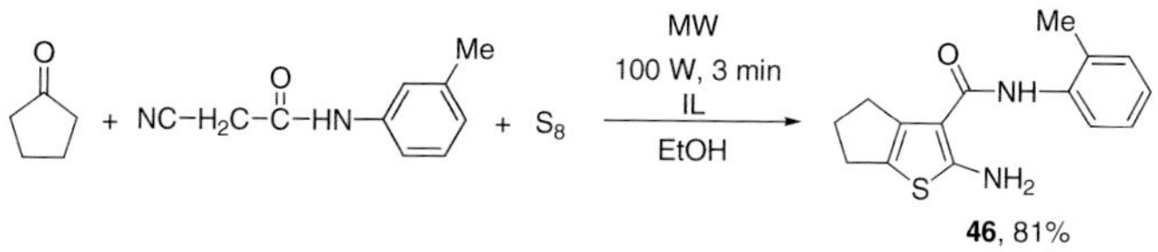

**Scheme 2.37**  Condensation of cyclopentanone, 2-cyano-*N*-*o*-tolylacetamide and sulfur

The synthesis of thiophene derivatives (**49**) on a soluble polymer-support utilizing the Gewald reaction was investigated (Scheme 2.38) [65]. The condensations were carried out in a household MW oven starting from various aldehydes or ketones, a PEG-supported cyanoacetic ester (**47**) and sulfur, in the presence of diisopropylethylamine (DIPEA) under solvent-free conditions. Then, the product (**48**) was acylated, and the PEG support was cleaved from the molecule by KCN in methanol. The desired thiophene derivatives (**49**) were obtained in yields of 48–95 %.

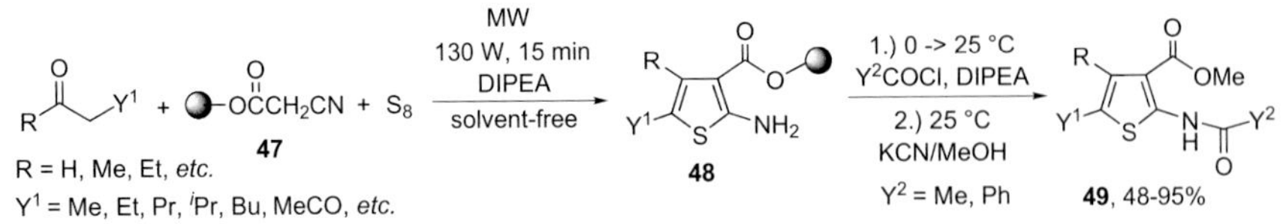

**Scheme 2.38** Synthesis of thiophene derivatives on a soluble polymer-support

## 2.3.10 Kabachnik-Fields Reactions

The Kabachnik-Fields reaction [66, 67] is a three-component condensation of an amine, an oxo compound, and a >P(O)H reagent forming α-aminophosphonates or α-aminophosphine oxides, which are synthetic targets of some importance, as the resulting species are the P–analogues of α-amino acids. See also Sect. 3.10.

A MW-assisted catalyst-free and solvent-free Kabachnik-Fields reaction of amines, aldehydes and dimethyl phosphite was described by Chinese researchers (Scheme 2.39) [68]. The condensations were carried out in a multimode MW reactor at 80 °C for 2 min, and the corresponding α-aminophosphonates (**50**) were obtained in yields of 40–98 %.

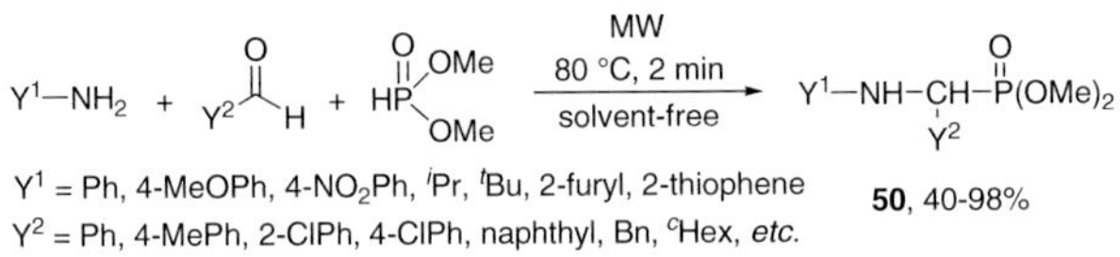

**Scheme 2.39** MW-assisted catalyst- and solvent-free Kabachnik-Fields reaction

Ordónez and co-workers reported a MW-assisted highly diastereoselective synthesis of α–aminophosphonates (**51**) by the three-component reactions of chiral amines, alkyl or aryl aldehydes and dimethyl phosphite (Scheme 2.40) [69]. The condensations were performed in the absence of any catalyst and solvent.

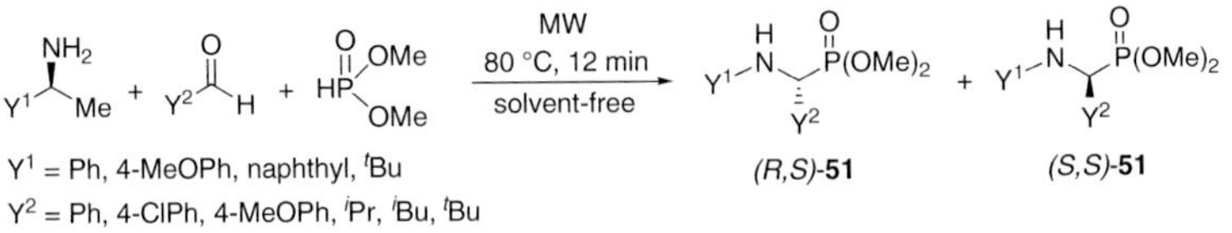

**Scheme 2.40** MW-assisted diastereoselective synthesis of α–aminophosphonates

Bis($\alpha$-aminophosphonate) pesticides were synthesized by the Kabachnik-Fields reaction of terephthalaldehyde, 2 equivalents of aniline derivatives and diethyl- or dibutyl phosphite under catalyst- and solvent-free MW-assisted conditions (Scheme 2.41) [70].

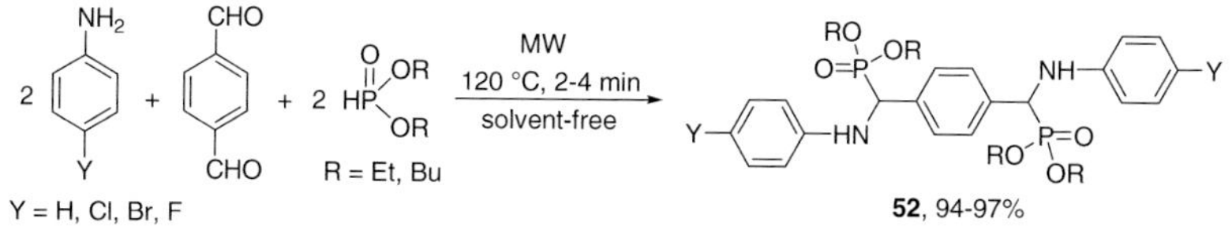

**Scheme 2.41**  Synthesis of bis($\alpha$-aminophosphonate) pesticides under MW conditions

## 2.3.11  Kindler Reactions

The Kindler reaction [71] is a three-component condensation of an aldehyde, an amine and elemental sulfur resulting in the formation of thioamides. The modification of this condensation, where ammonium polysulfide is used instead of sulfur, is the Willgerodt-Kindler reaction.

Thiobenzamide derivatives (**53**) were synthesized by the condensation of benzaldehyde or 4-(dimethylamino)benzaldehyde, morpholine and sulfur in a household MW oven (Scheme 2.42) [72]. The reactions were studied using acid and also base catalysts, and it was observed that the bases were more efficient.

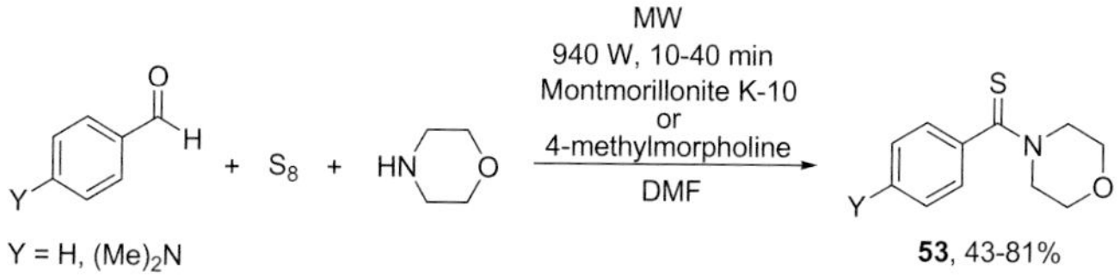

**Scheme 2.42**  MW-assisted synthesis of thiobenzamide derivatives

Another MW-assisted Kindler reaction was reported by Kappe and co-workers (Scheme 2.43) [73]. The three-component condensation of aldehydes, amines and elemental sulfur leading to thioamides **54** was performed using 1–methyl-2-pyrrolidone (NMP) as the solvent at 110–180 °C for 2–20 min.

**Scheme 2.43**  Kindler reaction of aldehydes, amines and elemental sulfur under MW irradiation

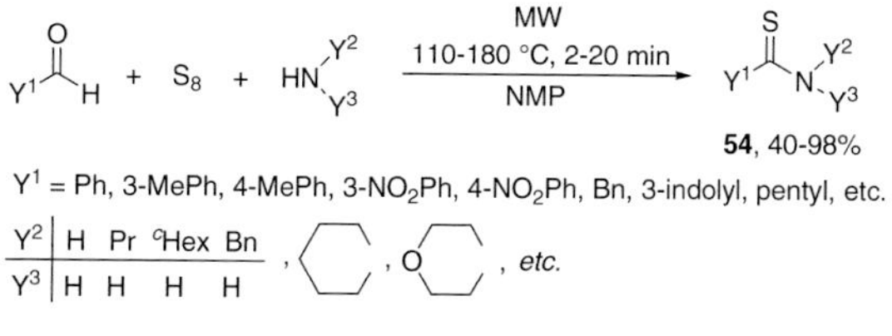

$Y^1$ = Ph, 3-MePh, 4-MePh, 3-NO$_2$Ph, 4-NO$_2$Ph, Bn, 3-indolyl, pentyl, etc.

| $Y^2$ | H | Pr | $^c$Hex | Bn |
|---|---|---|---|---|
| $Y^3$ | H | H | H | H |

### 2.3.12 Passerini Reactions

The Passerini reaction [74] is a multi-component transformation among a carboxylic acid, a ketone or an aldehyde, and an isocyanide to form the corresponding α-hydroxy carboxamide.

Brazilian researchers described the solvent-free MW-assited Passerini reaction of substituted carboxylic acids, aldehydes and isonitriles (Scheme 2.44) [75]. The corresponding α-acyloxy carboxamides (**55**) were obtained in good yields at 60 or 120 °C within 1–5 min.

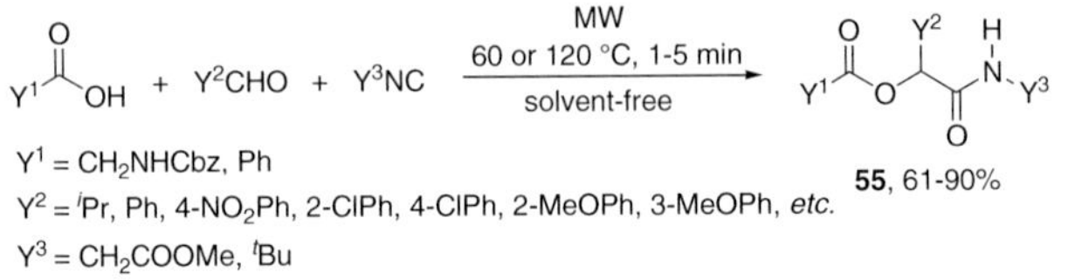
MW
60 or 120 °C, 1-5 min
solvent-free
55, 61-90%
Y1 = CH2NHCbz, Ph
Y2 = iPr, Ph, 4-NO2Ph, 2-ClPh, 4-ClPh, 2-MeOPh, 3-MeOPh, etc.
Y3 = CH2COOMe, tBu

**Scheme 2.44** Solvent-free MW-assited Passerini reaction of carboxylic acids, aldehydes and isonitriles

Boron-containing α-acyloxyamide analogues (**57** and **59**) were synthesized from a boron-containing acid (**56**), aldehydes and cyclohexyl isocyanide (Scheme 2.45 (1)), or from a boron-containing aldehyde (**58**), acids and cyclohexyl isocyanide (Scheme 2.45 (2)) in water under MW conditions [76].

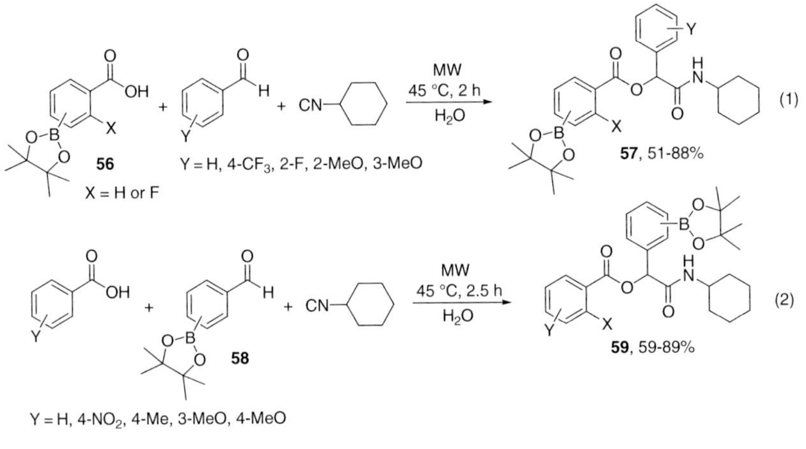
MW
45 °C, 2 h
H2O
56
X = H or F
Y = H, 4-CF3, 2-F, 2-MeO, 3-MeO
57, 51-88%
(1)
MW
45 °C, 2.5 h
H2O
58
59, 59-89%
(2)
Y = H, 4-NO2, 4-Me, 3-MeO, 4-MeO

**Scheme 2.45** The synthesis of boron-containing α-acyloxyamide analogues

The three-component reaction of trolox derivatives (**60**), furoxan aldehyde (**61**) and phenylethylisocyanide was also described (Scheme 2.46) [77]. The reactions were carried out in water at 60 °C under MW irradiation for 5 min.

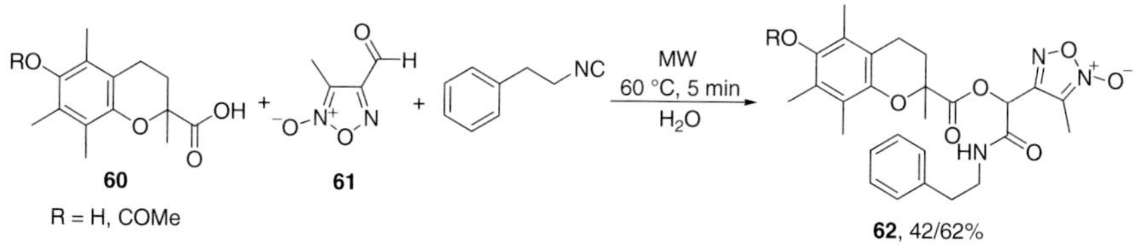

**Scheme 2.46** The condensation of trolox derivatives, furoxan aldehyde and phenylethylisocyanide

## 2.3.13 Ugi Reactions

The Ugi four-component condensation [78] with the participation of an amine, an aldehyde or ketone, a carboxylic acid and an isocyanide affords α-aminoacyl amide derivatives, which may be of potential pharmaceutical applications.

A one-pot Ugi reaction followed by intramolecular O–alkylation is an elegant example. The synthesis starts from 2-aminophenols, aldehydes, α–bromocarboxylic acids and isocyanides under MW irradiation (Scheme 2.47) [79].

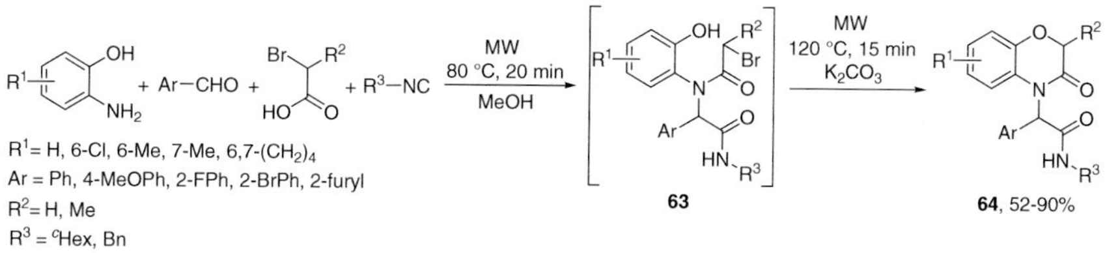

**Scheme 2.47** MW-assisted one-pot Ugi reaction followed by an intramolecular O–alkylation

The MW-assisted special Ugi reaction of levulinic acid, amines and isonitriles afforded the corresponding lactams (**65**) in moderate to excellent yields at 100 °C after 30 min (Scheme 2.48) [80].

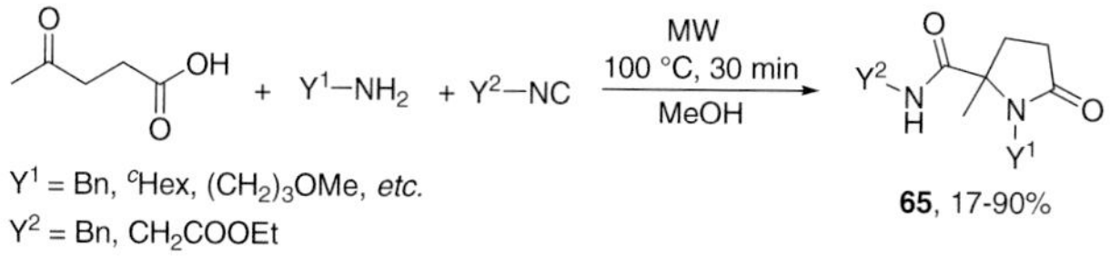

**Scheme 2.48** Ugi reaction of levulinic acid, amines and isonitriles under MW conditions

The synthesis of five- and six-membered lactams via Ugi reaction was also reported (Scheme 2.49) [81]. The condensation of 4-acetylbutyric acid or levulinic acid, amines and isocyanides was carried out under solvent-free MW conditions in a short time.

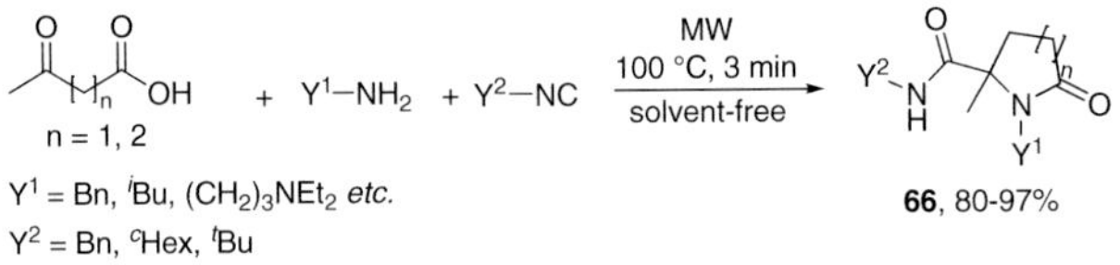
MW
100 °C, 3 min
solvent-free
n = 1, 2
Y¹ = Bn, ⁱBu, (CH₂)₃NEt₂ etc.
Y² = Bn, ᶜHex, ᵗBu
66, 80-97%

**Scheme 2.49** The synthesis of five- and six-membered lactams via MW-assisted Ugi reaction

## 2.3.14 *Domino Reactions*

In the domino reaction, called also tandem or cascade reaction, two or more transformations take place under the conditions applied without adding any additional reagents or catalysts. These reactions may include multistep synthesis and among others, protection-deprotection steps. Work-up procedures and purifications can be avoided.

Efficient four- and six-component domino reactions were developed, where 2–(2′–azaaryl)imidazoles (**67**) and *anti*-1,2-diarylethylbenzamide derivatives (**68**) were obtained under solvent-free MW-assisted conditions (Scheme 2.50 (1) and (2)) [82].

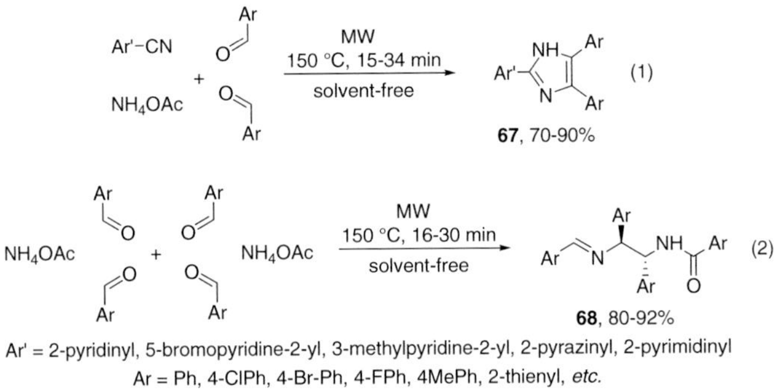
Ar'-CN
NH₄OAc
MW
150 °C, 15-34 min
solvent-free
(1)
67, 70-90%
NH₄OAc
NH₄OAc
MW
150 °C, 16-30 min
solvent-free
(2)
68, 80-92%
Ar' = 2-pyridinyl, 5-bromopyridine-2-yl, 3-methylpyridine-2-yl, 2-pyrazinyl, 2-pyrimidinyl
Ar = Ph, 4-ClPh, 4-Br-Ph, 4-FPh, 4MePh, 2-thienyl, etc.

**Scheme 2.50** MW-assisted solvent-free four- and six-component domino reactions

Substituted quinolones (**69**) were prepared by a montmorillonite K-10 catalyzed multicomponent domino reaction of amines, aldehydes and terminal arylalkyne under MW irradiation (Scheme 2.51) [83].

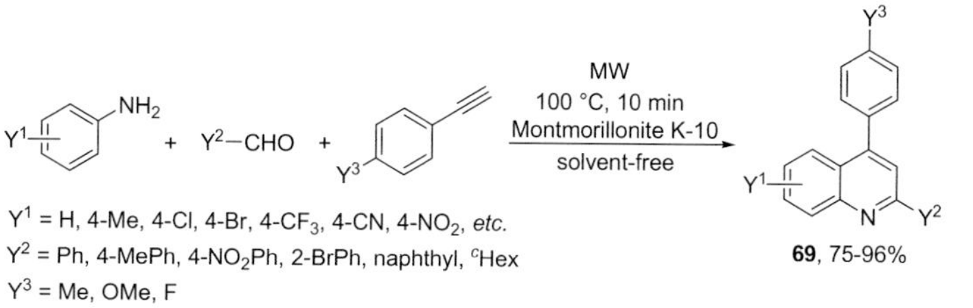
MW
100 °C, 10 min
Montmorillonite K-10
solvent-free
NH2
Y1
Y2–CHO
Y3
Y2
N
69, 75-96%
Y1 = H, 4-Me, 4-Cl, 4-Br, 4-CF3, 4-CN, 4-NO2, etc.
Y2 = Ph, 4-MePh, 4-NO2Ph, 2-BrPh, naphthyl, cHex
Y3 = Me, OMe, F

**Scheme 2.51** Montmorillonite K-10 catalyzed multicomponent domino reactions

## 2.4 Cycloadditions

Cycloaddition reactions are pericyclic reactions in which two or more unsaturated compounds are combined with the formation of a cyclic adduct. Thus, cycloadditions provide heterocyclic and multicyclic scaffolds in a single-step. Cycloadditions involving atomic efficient transformations represent another widely investigated group of MW-assisted organic reactions [84].

### *2.4.1 [2+2] Cycloadditions*

[2+2] Cycloadditions provide a synthetic tool towards four member rings, such as cyclobutanes, cyclobutenes, β-lactams, oxetenes, cyclobutanones, and their derivatives. These reactions usually require photochemical activation, or the use of a Lewis acid under thermal conditions. A few examples were described, where MW irradiation was found to be beneficial [85].

Ovaska reported a facile MW-assisted intramolecular [2+2] cycloaddition starting from germinal allenyl-propargyl-substituted cyclopentane derivatives (**70**), leading to strained tricyclic 5–6–4 ring systems (**71**) resembling to natural sterpurenes (Scheme 2.52) [86].

**Scheme 2.52** The formation of strained tricyclic 5–6–4 ring systems

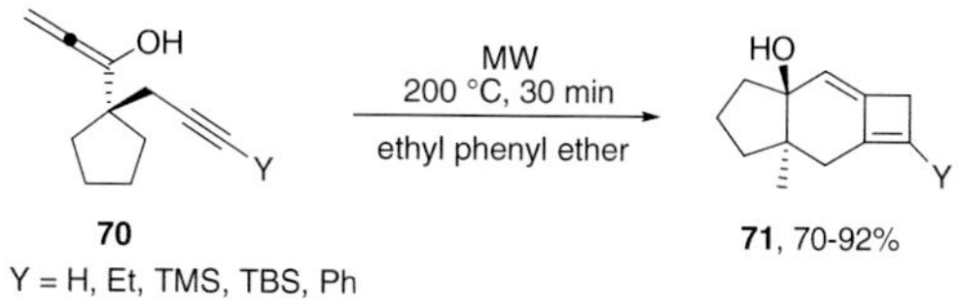
OH
Y
MW
200 °C, 30 min
ethyl phenyl ether
HO
Y
70
Y = H, Et, TMS, TBS, Ph
71, 70-92%

A similar regioselective intramolecular cycloaddition was described for the formation of bicyclic compounds by Brummond and co-workers. Bicycloalkadienes were formed efficiently when **72** was irradiated by MW at 250 °C in toluene, in the presence of an IL as an additive (Scheme 2.53) [87].

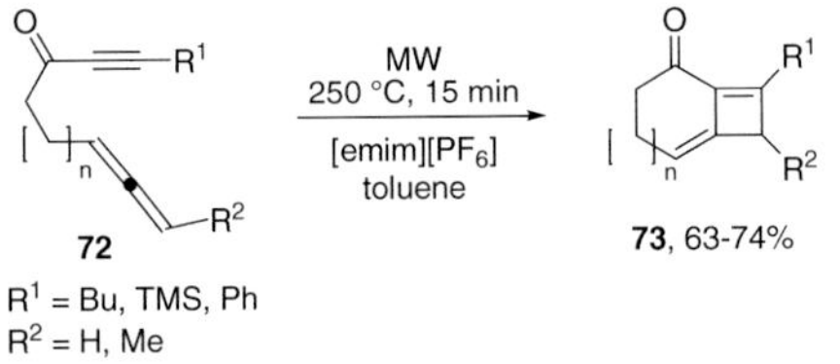
MW
250 °C, 15 min
[emim][PF6]
toluene
72
R1 = Bu, TMS, Ph
R2 = H, Me
73, 63-74%

**Scheme 2.53** MW-assisted regioselective intramolecular cycloaddition leading to bicycloalkadienes

An intermolecular Staudinger [2+2] cycloaddition of a phenyl thiodiazoacetate to an imine was enhanced by MW irradiation (Scheme 2.54). The authors aimed at the investigation of nonthermal microwave effects (see also Chap. 4.), and they found no significant difference in the stereoselectivity of the MW-assisted or the thermal variation. However, the reaction speed was somewhat increased under MW conditions [88].

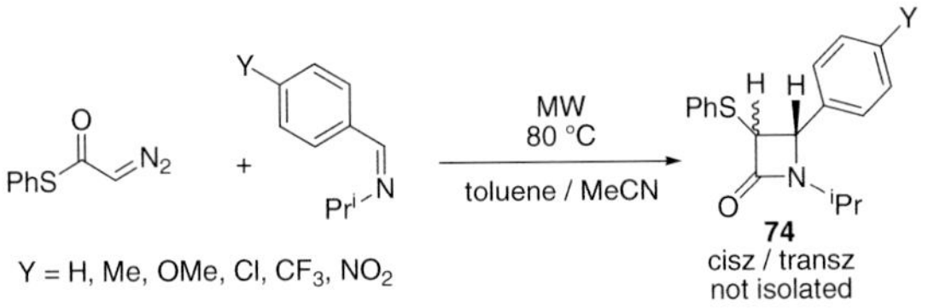
MW
80 °C
toluene / MeCN
Y = H, Me, OMe, Cl, CF3, NO2
74
cisz / transz
not isolated

**Scheme 2.54** A MW-enhanced intermolecular Staudinger cycloaddition

A regiospecific protocol was described for the formation of highly functionalized dienes. 2–Amino-3-dimethylaminopropenoates were reacted with acetylene derivatives to furnish eventually 1–amino-4-(dimethylamino)buta-1,3-diene derivatives (**76**) (Scheme 2.55). The reaction takes place via a cyclobutene intermediate (**75**) by retro-electrocyclisation [89].

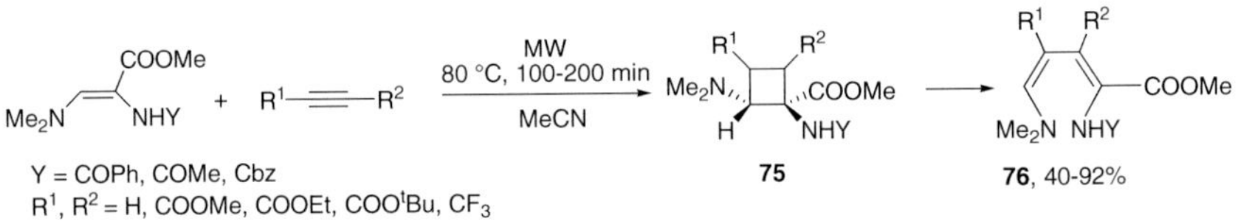
MW
80 °C, 100-200 min
MeCN
Y = COPh, COMe, Cbz
R1, R2 = H, COOMe, COOEt, COOtBu, CF3
75
76, 40-92%

**Scheme 2.55** A regiospecific [2+2] cycloaddition leading to buta-1,3-diene derivatives

Pfeffer and co-workers investigated the synthesis of dicyclobutene tetraester **77** by the reaction of norbornadiene and DMAD in the presence of $[RuH_2(CO)(PPh_3)_3]$ as a catalyst complex. While under conventional heating almost no product formation was obtained, under MW irradiation, the corresponding cycloadduct was formed already after 2 min (Scheme 2.56) [90].

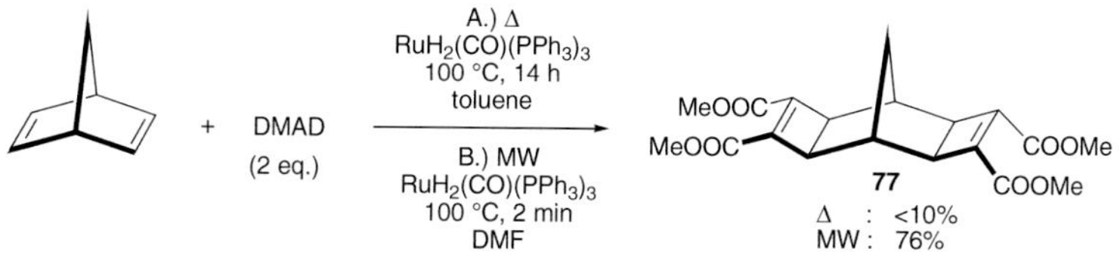

**Scheme 2.56** MW-assisted synthesis of a dicyclobutene tetraester

## 2.4.2 [3+2] Cycloadditions

1,3-Dipolar cycloadditions are among the most efficient procedures to form five-membered heterocycles [84]. The reaction of azides with alkynes or nitriles are powerful "click reactions" resulting in 1,2,3-triazoles or tetrazoles. Under traditional thermal conditions, these cycloadditions require often high reaction temperatures.

The copper-catalyzed azide–alkyne cycloaddition (CuAAC) is one of the best "click reactions" to date, as the use of Cu(I) catalysts provides a significant rate acceleration as compared to the uncatalyzed 1,3-dipolar cycloaddition [91]. Several examples confirmed that further enhancement can be obtained by MW irradiation [92].

A new green method have been developed for the formation of 1,2,3-triazoles by Taher and co-workers. A highly active and stable poly-phenylenediamine supported copper(I) catalyst (Cu(I)-pPDA) was found to promote the 1,3-dipolar cycloaddition between terminal alkynes and azides (Scheme 2.57). Thus, the MW-assisted solvent-free accomplishment provides 1,2,3-triazoles (**74**) of pharmaceutical importance with excellent yields [93].

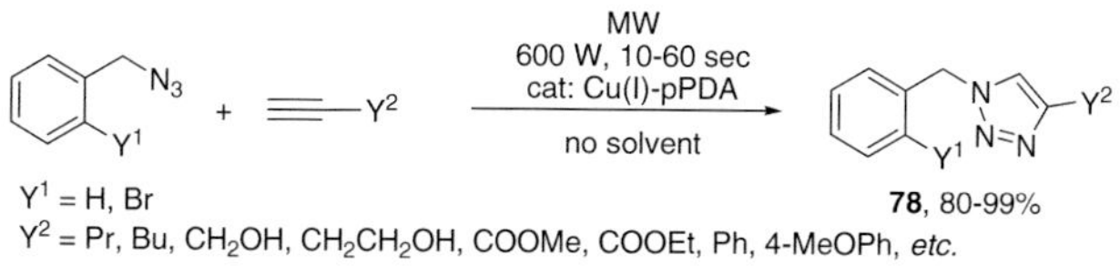

$Y^1$ = H, Br
$Y^2$ = Pr, Bu, $CH_2OH$, $CH_2CH_2OH$, COOMe, COOEt, Ph, 4-MeOPh, etc.

**Scheme 2.57** An environmentally benign synthesis of 1,2,3-triazoles

Other metal-catalyzed azide–alkyne cycloaddition reactions have also been reported under MW heating. The synthesis of 1,2,3-triazoles via Ru-catalyzed azide–alkyne cycloaddition (RuAACs) was described by Fokin. It is noteworthy that while the 1,4-disubstituted triazoles were obtained in the Cu(I)-catalyzed azide–alkyne cycloaddition, the Ru-catalyzed version led to the 1,5-regioisomers of 1,2,3-triazoles (Scheme 2.58). MW irradiation provided higher yields, cleaner products in shorter reaction times, as compared to the results obtained on traditional heating, upon which by-products were also formed [94].

**Scheme 2.58** MW-assisted Ru-catalyzed azide–alkyne cycloaddition

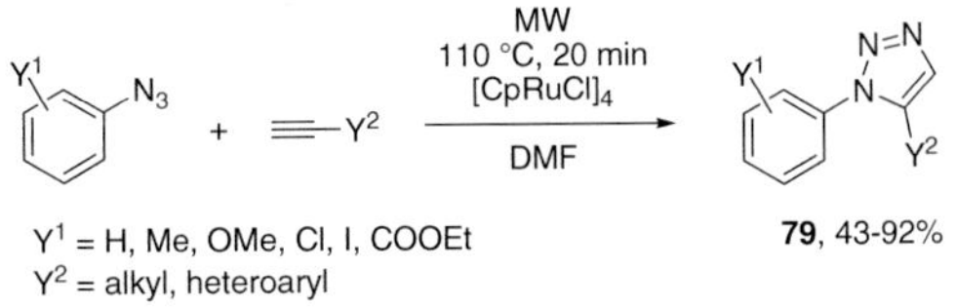
Y1
N3
+
Y2
MW
110 °C, 20 min
[CpRuCl]4
DMF
Y1
N=N
N
Y2
Y1 = H, Me, OMe, Cl, I, COOEt
Y2 = alkyl, heteroaryl
79, 43-92%

An interesting example for 1,3-dipolar cycloadditions is the reaction of an azaphosphonate and an acetylenic ester to furnish the corresponding 1,2,3-triazole as a mixture of two regioisomers (**80A** and **80B**) (Scheme 2.59). While the reaction took place in toluene at 110 °C after 30 h, the solvent-free MW-assisted variation was complete after 5 min [95].

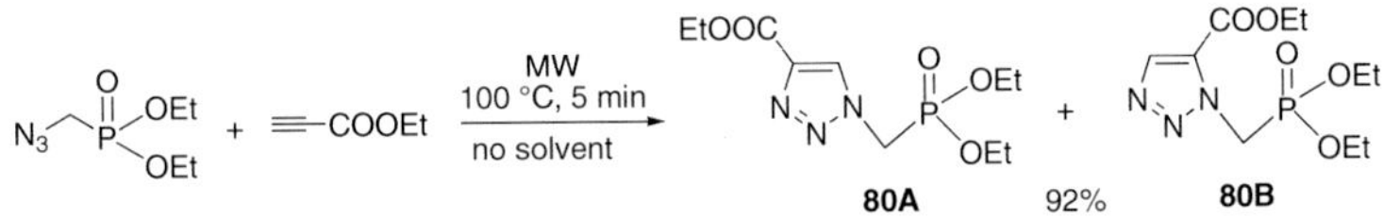
N3
O
P
OEt
OEt
+
COOEt
MW
100 °C, 5 min
no solvent
EtOOC
N
N
N
O
P
OEt
OEt
+
COOEt
80A
92%
80B

**Scheme 2.59** 1,3-Dipolar cycloaddition of an azaphosphonate to an acetylenic ester

Kappe described the first example of an organocatalytic tetrazole-formation under MW-assisted conditions. The catalyst (5-azido-1-methyl-3,4-dihydro-2*H*-pyrrolium azide) was formed in situ. The cycloaddition of azides with organic nitriles resulted in a series of 5-substituted-1*H*-tetrazoles in high yields (Scheme 2.60) [96].

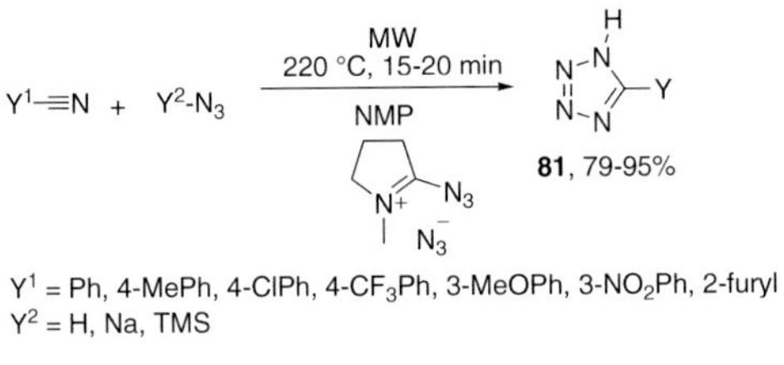
Y1≡N + Y2-N3
MW
220 °C, 15-20 min
NMP
H
N
N
N
N
Y
81, 79-95%
N+
N3
N3
Y1 = Ph, 4-MePh, 4-ClPh, 4-CF3Ph, 3-MeOPh, 3-NO2Ph, 2-furyl
Y2 = H, Na, TMS

**Scheme 2.60** Organocatalytic tetrazole-formation under MW irradiation

### 2.4.3 Diels–Alder Cycloadditions

The [4+2] cycloaddition of a conjugated diene and a dienophile is widely used to form highly functionalized and fused ring systems. In most cases, the syntheses take place with a high degree of chemo-, regio- and stereoselectivity.

Triazoles are known for their poor reactivity in [4+2] cycloaddition reactions. However, an example was described in which the 1,2,3-triazole ring acted as a diene towards dimethyl acetylenedicarboxylate (DMAD) in MW-assisted solvent-free Diels–Alder cycloadditions followed by a rearrangement to afford functionalized pyrazole heterocycles (**83**) (Scheme 2.61). The yields could be increased using a supported Lewis acid catalyst, which could be recycled at least five times without a decrease of activation [97].

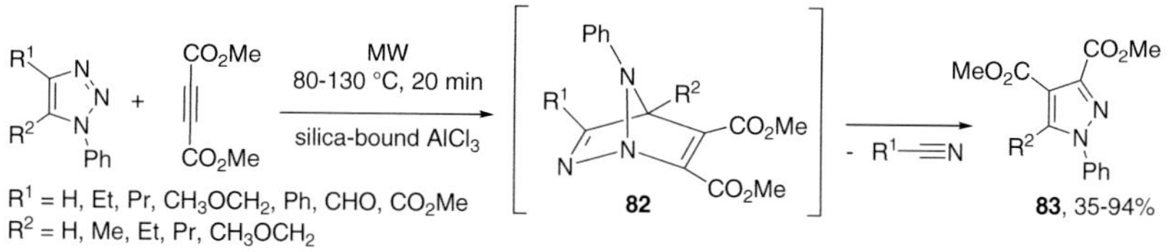

$R^1$ = H, Et, Pr, CH$_3$OCH$_2$, Ph, CHO, CO$_2$Me
$R^2$ = H, Me, Et, Pr, CH$_3$OCH$_2$

**Scheme 2.61** MW-assisted solvent-free [4+2] cycloaddition of triazoles to DMAD

Zheng observed the Diels–Alder reaction between Danishefsky's diene and ethyl α-substituted acrylate derivatives to provide cycloadducts **85** (after deprotection with (+)-10-camphorsulfonic acid (CSA) or pyridinium p-toluene sulfonate (PPTS) from **84**) (Scheme 2.62). The MW heating drastically accelerated the cycloaddition resulting in the desired products in high yields. Compared to the traditional thermal conditions, the method of Zheng offers a 14–48-fold rate acceleration with serious increase in the yields. The adducts so-obtained are useful intermediates in the synthesis of a biotin conjugate of monocyclic cyanoenone with high antiinflammatory activity [98].

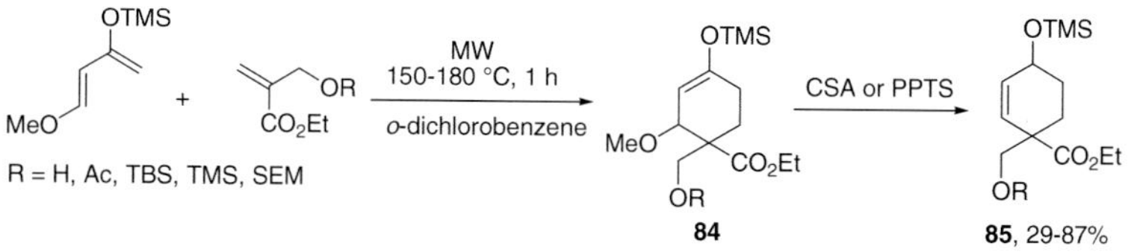

R = H, Ac, TBS, TMS, SEM

**Scheme 2.62** An effective MW-assisted [4+2] cycloaddition of Danishefsky's diene

The MW-assisted intramolecular Diels–Alder cyclization of alkenylaminofuranes at 180 °C in o-dichlorobenzene led to 4-monosubstituted indoles (**87**) after dehydrative aromatization of intermediate **86** (Scheme 2.63). Interestingly, no reaction was observed on conventional heating, whereas under MW-assisted conditions, the cyclization furnished the desired 4-substituted indoles in high yields [99]. Thus, the strategy shown is a convenient alternative to the transition metal-mediated coupling processes affording such heterocycles.

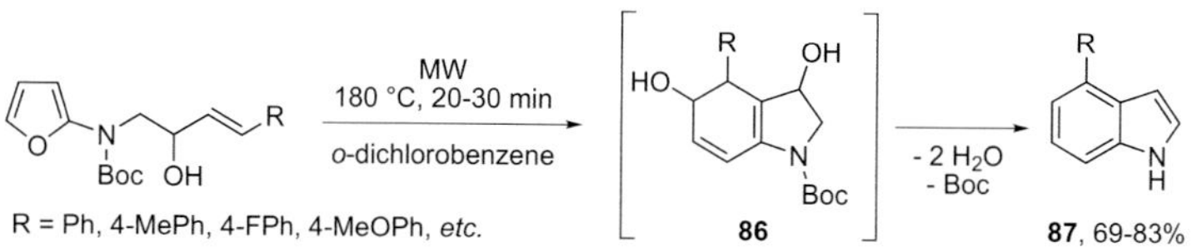

R = Ph, 4-MePh, 4-FPh, 4-MeOPh, etc.

**Scheme 2.63** Intramolecular Diels–Alder cyclization of furan derivatives

Kočevar developed an efficient synthesis of 1,5,6-trisubstituted indoles involving two MW-assisted steps. The first step is the Diels–Alder cycloaddition reaction between (*Z*)-1-methoxybut-1-en-3-yne with 2*H*-pyran-2-ones (**88**) yielding substituted aniline derivatives (**89**). In the next step, the adducts underwent intramolecular cyclization under acidic conditions to give the corresponding indole derivatives (**90**) (Scheme 2.64). It is worth mentioning that the analogous cycloaddition reactions carried out under high-pressure conventional heating conditions needed very long reaction times up to 138 days, and in two cases anomalous products were obtained [100].

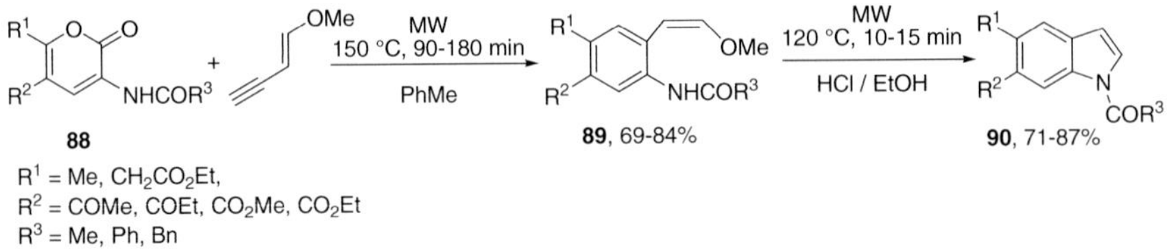

MW
150 °C, 90-180 min
PhMe
MW
120 °C, 10-15 min
HCl / EtOH
88
89, 69-84%
90, 71-87%
R1 = Me, CH2CO2Et,
R2 = COMe, COEt, CO2Me, CO2Et
R3 = Me, Ph, Bn


**Scheme 2.64** MW-assisted synthesis of 1,5,6-trisubstituted indoles

The Diels–Alder cycloaddition of 3-nitro-1-(p-toluenesulfonyl)pyrrole with *N*-acetyl-*N*-isopropyl-1,3-butadiene afforded an indole derivative (**91**) under solvent-free MW-assisted conditions after the elimination of the nitro group and in situ aromatization [101]. It is noted that the reaction did not occur on conventional heating (Scheme 2.65).

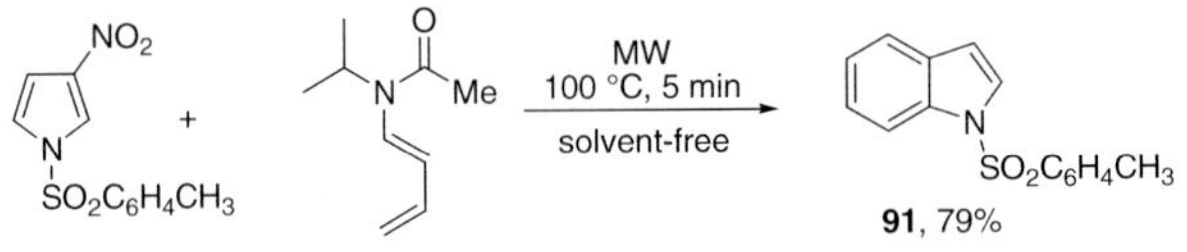

MW
100 °C, 5 min
solvent-free
91, 79%


**Scheme 2.65** Solvent-free MW-assisted [4+2] cycloaddition of a pyrrole derivative

A MW-assisted intramolecular didehydrogenative Diels–Alder reaction of styrene-ynes (**92**) was reported to furnish fluorophores **93** (Scheme 2.66) [102, 103].

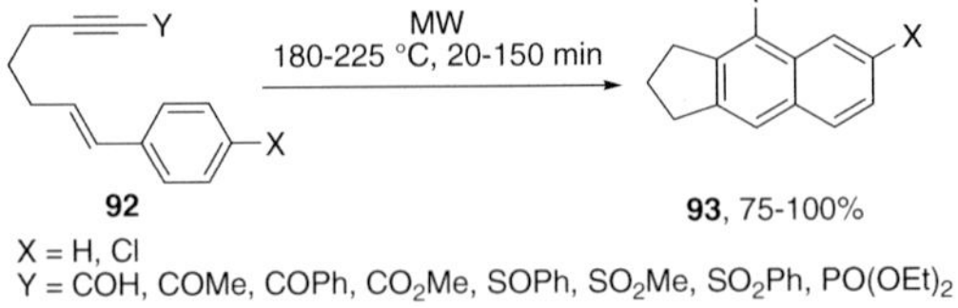

MW
180-225 °C, 20-150 min
92
93, 75-100%
X = H, Cl
Y = COH, COMe, COPh, CO2Me, SOPh, SO2Me, SO2Ph, PO(OEt)2


**Scheme 2.66** Didehydrogenative Diels–Alder reaction of styrene-ynes under MW conditions

4-Substituted-2,3-dihydrofuro[2,3-b]pyridines and 5-substituted-3,4-dihydro-2H-pyrano[2,3-b]pyridines (95) featuring close structural similarity to bioactive molecules were obtained by the intramolecular hetero Diels–Alder cycloaddition of alkyne triazines (94) under MW conditions in good yields (Scheme 2.67) [104]. MW activation proved to be efficient to promote the cycloaddition reaction.

Ar = pyrid-4-yl, thien-2-yl, 4-MePh, 4-NO$_2$Ph, 4-MeOPh

**Scheme 2.67** Intramolecular inverse electron demand Diels–Alder reactions under MW irradiation

1,4-Dihydropyridines (97) were prepared by an aza-Diels–Alder [4+2] cycloaddition strategy (Scheme 2.68) promoted by MW irradiation. The 1,4-dihydropyridine prepared (97) was converted further to antihypertensive drug Amlodipine (not shown here) [105].

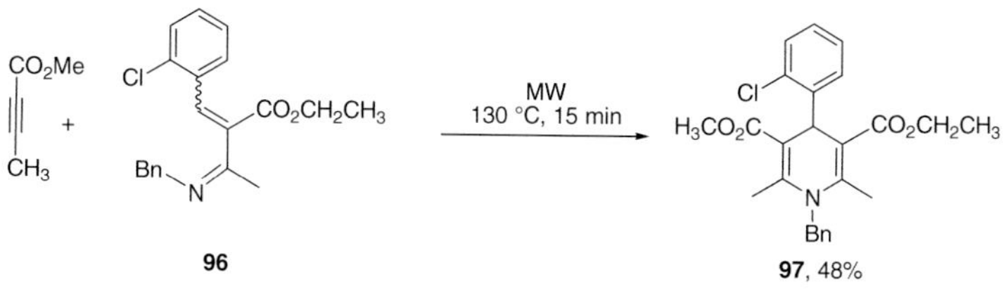

**Scheme 2.68** Aza-Diels–Alder cycloaddition to form an 1,4-dihydropyridine derivative

The MW-assisted [4+2] cycloadditions for the synthesis of drug-like heterocycles was also reported. The [4+2] cycloaddition of 1,4-diaryl-1-aza-1,3-butadienes (98) with allenic esters at 100 °C followed by a tandem 1,3-H-shift provided 1,4-dihydropyridines (100) in excellent, 83–96 % yields (Scheme 2.69). Comparative thermal reactions required 33–76 h resulting in lower yields [106]. The unsymmetrically substituted 1,4-dihydropyridines (100) obtained are well-known for their potential biological activities.

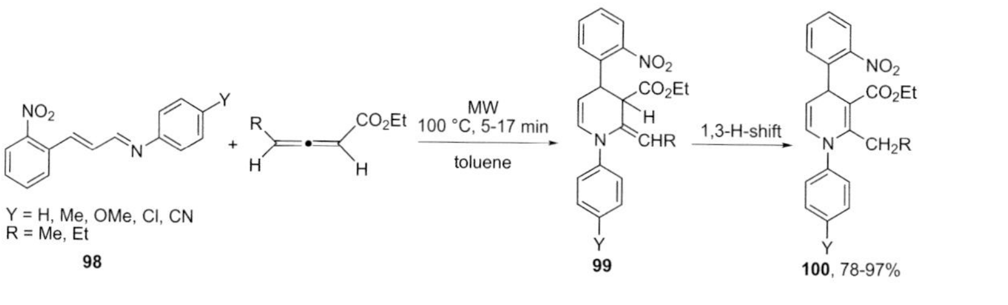

Y = H, Me, OMe, Cl, CN
R = Me, Et

**Scheme 2.69** MW-assisted synthesis of unsymmetrically substituted 1,4-dihydropyridines

## 2.5  Conclusions

In summary, MW-assisted coupling reactions, condensations, multicomponent reactions and cycloadditions providing an access to a wide variety of different scaffolds were presented. In all cases, MW irradiation led to shorter reaction times and higher yields in comparison with conventional heating, or even promoted reactions that were unsuccessful on conventional conditions.

## References

1. Gupta AK, Singh N, Singh KN (2015) On water synthesis of highly functionalized 4Hchromenes via carbon–carbon bond formation under microwave irradiation and their antibacterial properties. Curr Org Chem 5:28958–28964. doi:10.1039/c5ra01301a
2. Allam BK, Singh KN (2011) An efficient phosphine-free heck reaction in water using Pd (1-proline)2 as the catalyst under microwave irradiation. Synthesis 2011:1125–1131. doi:10.1055/s-0030-1258452
3. Herve G, Len C (2014) First ligand-free, microwave-assisted, Heck cross-coupling reaction in pure water on a nucleoside—application to the synthesis of antiviral BVDU. RSC Adv 4:46926–46929. doi:10.1039/C4RA09798J
4. Len C, Hervé G (2015) Aqueous microwaves assisted cross-coupling reactions applied to unprotected nucleosides. Front Chem 3:10. doi:10.3389/fchem.2015.00010
5. Dighe MG, Degani MS (2011) Microwave-assisted ligand-free, base-free Heck reactions in a task-specific imidazolium ionic liquid. Arkivoc XI:189–197. doi:10.3998/ark.5550190.0012.b17
6. Bai L, Wang J-X, Zhang Y (2003) Rapid microwave-promoted Suzuki cross coupling reaction in water. Green Chem 5:615–617. doi:10.1039/B305191A
7. Crozet MD, Castera-Ducros C, Vanelle P (2006) An efficient microwave-assisted Suzuki cross-coupling reaction of imidazo[1,2-a]pyridines in aqueous medium. Tetrahedron Lett 47:7061–7065. doi:10.1016/j.tetlet.2006.07.098
8. Dawood KM, El-Deftar MM (2010) Microwave-assisted C-C cross-coupling reactions of aryl and heteroaryl halides in water. Arkivoc 9:319–330. doi:10.3998/ark.5550190.0011.930
9. Cohen A, Crozet MD, Rathelot P, Vanelle P (2009) An efficient aqueous microwave-assisted Suzuki-Miyaura cross-coupling reaction in the thiazole series. Green Chem 11:1736–1742. doi:10.1039/B916123F
10. Saha P, Naskar S, Paira P, Hazra A, Sahu KB, Paira R, Banerjee S, Mondal NB (2009) Basic alumina-supported highly effective Suzuki-Miyaura cross-coupling reaction under microwave irradiation: application to fused tricyclic oxa-aza-quinolones. Green Chem 11:931–934. doi:10.1039/B902916H
11. Baghbanzadeh M, Pilger C, Kappe CO (2011) Rapid nickel-catalyzed Suzuki–Miyaura cross-couplings of aryl carbamates and sulfamates utilizing microwave heating. J Org Chem 76:1507–1510. doi:10.1021/jo1024464
12. Yılmaz Ü, Küçükbay H, Şireci N, Akkurt M, Günal S, Durmaz R, Nawaz Tahir M (2011) Synthesis, microwave-promoted catalytic activity in Suzuki-Miyaura cross-coupling reactions and antimicrobial properties of novel benzimidazole salts bearing trimethylsilyl group. Appl Organomet Chem 25:366–373. doi:10.1002/aoc.1772
13. Gupta AK, Singh N, Singh KN (2013) Microwave assisted organic synthesis: cross coupling and multicomponent reactions. Curr Org Chem 17:474–490. doi:10.2174/1385272811317050005

14. Shah D, Kaur H (2012) Macroporous resin impregnated palladium nanoparticles: Catalyst for a microwave-assisted green Hiyama reaction. J Mol Catal A: Chem 359:69–73. doi:10.1016/j.molcata.2012.03.022
15. Gupta AK, Tirumaleswara Rao G, Singh KN (2012) NiCl2·6H2O as recyclable heterogeneous catalyst for N-arylation of amines and NH-heterocycles under microwave exposure. Tetrahedron Lett 53:2218–2221. doi:10.1016/j.tetlet.2012.02.081
16. Guo D, Huang H, Zhou Y, Xu J, Jiang H, Chen K, Liu H (2010) Ligand-free iron/copper cocatalyzed N-arylations of aryl halides with amines under microwave irradiation. Green Chem 12:276–281. doi:10.1039/B917010C
17. Liu Z-J, Vors J-P, Gesing ERF, Bolm C (2011) Microwave-assisted solvent- and ligand-free copper-catalysed cross-coupling between halopyridines and nitrogen nucleophiles. Green Chem 13:42–45. doi:10.1039/C0GC00296H
18. Chow WS, Chan TH (2009) Microwave-assisted solvent-free N-arylation of imidazole and pyrazole. Tetrahedron Lett 50:1286–1289. doi:10.1016/j.tetlet.2008.12.119
19. Hirao T, Masunaga T, Ohshiro Y, Agawa T (1980) Stereoselective synthesis of vinylphosphonate. Tetrahedron Lett 21:3595–3598. doi:10.1016/0040-4039(80)80245-0
20. Kalek M, Ziadi A, Stawinski J (2008) Microwave-assisted palladium-catalyzed cross-coupling of aryl and vinyl halides with H-phosphonate diesters. Org Lett 10:4637–4640. doi:10.1021/ol801935r
21. Jiang W, Allan G, Fiordeliso JJ, Linton O, Tannenbaum P, Xu J, Zhu P, Gunnet J, Demarest K, Lundeen S, Sui Z (2006) New progesterone receptor antagonists: phosphorus-containing 11β-aryl-substituted steroids. Bioorg Med Chem 14:6726–6732. doi:10.1016/j.bmc.2006.05.066
22. Andaloussi M, Lindh J, Sävmarker J, Sjöberg PJR, Larhed M (2009) Microwave-promoted palladium(II)-catalyzed C–P bond formation by using arylboronic acids or aryltrifluoroborates. Chem Eur J 15:13069–13074. doi:10.1002/chem.200901473
23. Julienne D, Lohier J-F, Delacroix O, Gaumont A-C (2007) Palladium-catalyzed C–P coupling reactions between vinyl triflates and phosphine–boranes: Efficient access to vinylphosphine–boranes. J Org Chem 72:2247–2250. doi:10.1021/jo062482o
24. Jablonkai E, Keglevich G (2013) P-ligand-free, microwave-assisted variation of the Hirao reaction under solvent-free conditions; the P-C coupling reaction of >P(O)H species and bromoarenes. Tetrahedron Lett 54:4185–4188. doi:10.1016/j.tetlet.2013.05.111
25. Jablonkai E, Keglevich G (2014) Advances and new variations of the Hirao reaction. Org Prep Proced Int 46:281–316. doi:10.1080/00304948.2014.922376
26. Kohler MC, Sokol JG, Stockland RA Jr (2009) Development of a room temperature Hirao reaction. Tetrahedron Lett 50:457–459. doi:10.1016/j.tetlet.2008.11.040
27. Bagley M, Lubinu MC (2006) Microwave-assisted multicomponent reactionsfor the synthesis of heterocycles. In: Van der Eycken E, Kappe CO (eds) Microwave-assisted synthesis of heterocycles, vol 1. Topics in Heterocyclic Chemistry. Springer, Berlin, pp 31–58. doi:10.1007/7081_004
28. Marjani K, Asgari M, Ashouri A, Mahdavinia GH, Ahangar HA (2009) Microwave-assisted aldol condensation of benzil with ketones. Chin Chem Lett 20:401–403. doi:10.1016/j.cclet.2008.12.036
29. Zhang C, Wang B-S, Chen S-F, Zhang S-Q, Cui D-M (2011) Synthesis of 3-methyl-2-cyclohexenones catalyzed by mercury(II) salts and their microwave assisted BiCl3 catalyzed aldol condensations. J Organomet Chem 696:165–169. doi:10.1016/j.jorganchem.2010.08.037
30. Claisen L, Claparède A (1881) Condensationen von Ketonen mit Aldehyden. Ber Dtsch Chem Ges 14:2460–2468. doi:10.1002/cber.188101402192
31. Min-yi L, Dan W, Zhi-xiong L (2009) Synthesis of tert-butyl-methoxy-dibenzoylmethane under microwave irradiation. Chin Surf Det Cosm 39:179–182
32. Knoevenagel E (1898) Condensation von Malonsäure mit aromatischen Aldehyden durch Ammoniak und Amine. Ber Dtsch Chem Ges 31:2596–2619. doi:10.1002/cber.18980310308

33. Ansari MB, Jin H, Parvin MN, Park S-E (2012) Mesoporous carbon nitride as a metal-free base catalyst in the microwave assisted Knoevenagel condensation of ethylcyanoacetate with aromatic aldehydes. Catal Today 185:211–216. doi:10.1016/j.cattod.2011.07.024

34. Sundaram K, Ravi S (2013) Microwave-assisted synthesis of 3-α-carboxyethylrhodanine derivatives and their in vitro antibacterial activity. J Appl Pharm Sci 3:133–135. doi:10.7324/JAPS.2013.3725

35. Biradar JS, Sasidhar BS (2011) Solvent-free, microwave assisted Knoevenagel condensation of novel 2,5-disubstituted indole analogues and their biological evaluation. Eur J Med Chem 46:6112–6118. doi:10.1016/j.ejmech.2011.10.004

36. Mannich C, Krösche W (1912) Ueber ein kondensationsprodukt aus formaldehyd, ammoniak und antipyrin. Arch Pharm 250:647–667. doi:10.1002/ardp.19122500151

37. Sankappa Rai U, Isloor AM, Shetty P, Isloor N, Malladi S, Fun H-K (2010) Synthesis and biological evaluation of aminoketones. Eur J Med Chem 45:6090–6094. doi:10.1016/j.ejmech.2010.09.015

38. Lehmann F, Pilotti Å, Luthman K (2003) Efficient large scale microwave assisted Mannich reactions using substituted acetophenones. Mol Divers 7:145–152. doi:10.1023/B:MODI.0000006809.48284.ed

39. Leadbeater N, Torenius H, Tye H (2003) Microwave-assisted Mannich-type three-component reactions. Mol Divers 7:135–144. doi:10.1023/B:MODI.0000006822.51884.e6

40. Dong X, Liu T, Chen J, Ying H, Hu Y (2009) Microwave-assisted Mannich reaction of 2-hydroxy-chalcones. Synth Commun 39:733–742. doi:10.1080/00397910802431107

41. Biginelli P (1891) Ueber aldehyduramide des acetessigäthers. Ber Dtsch Chem Ges 24:1317–1319. doi:10.1002/cber.189102401228

42. Kappe CO (2000) Biologically active dihydropyrimidones of the Biginelli-type—a literature survey. Eur J Med Chem 35:1043–1052. doi:10.1016/S0223-5234(00)01189-2

43. Kumar PBR, Sankar G, Nasir Baig RB, Chandrashekaran S (2009) Novel Biginelli dihydropyrimidines with potential anticancer activity: a parallel synthesis and CoMSIA study. Eur J Med Chem 44:4192–4198. doi:10.1016/j.ejmech.2009.05.014

44. Ranjith C, Srinivasan GV, Vijayan KK (2010) Tributyl borate mediated Biginelli reaction: a facile microwave-assisted green synthetic strategy toward dihydropyrimidinones. Bull Chem Soc Jpn 83:288–290. doi:10.1002/chin.201029176

45. Fu R, Yang Y, Lai W, Ma Y, Chen Z, Zhou J, Chai W, Wang Q, Yuan R (2015) Efficient and green microwave-assisted multicomponent Biginelli reaction for the synthesis of dihydropyrimidinones catalyzed by heteropolyanion-based ionic liquids under solvent-free conditions. Synth Commun 45:467–477. doi:10.1080/00397911.2014.976346

46. Virsodia VR, Vekariya NR, Manvar AT, Khunt RC, Marvania BR, Savalia BS, Shah AK (2008) Catalyst-free, rapid synthesis of fused bicyclic thiazolo-pyrimidine and pyrimido-thiazine derivatives by a microwave-assisted method. Phosphorus Sulfur Silicon Relat Elem 184:34–44. doi:10.1080/10426500802077564

47. Fang Z, Lam Y (2011) A rapid and convenient synthesis of 5-unsubstituted 3,4-dihydropyrimidin-2-ones and thiones. Tetrahedron 67:1294–1297. doi:10.1016/j.tet.2010.11.075

48. Legeay J-C, Vanden Eynde JJ, Bazureau JP (2005) Ionic liquid phase technology supported the three component synthesis of Hantzsch 1,4-dihydropyridines and Biginelli 3,4-dihydropyrimidin-2(1H)-ones under microwave dielectric heating. Tetrahedron 61:12386–12397. doi:10.1016/j.tet.2005.09.118

49. Hantzsch A (1881) Condensationsprodukte aus aldehydammoniak und ketonartigen verbindungen. Ber Dtsch Chem Ges 14:1637–1638. doi:10.1002/cber.18810140214

50. Öhberg L, Westman J (2001) An Efficient and fast procedure for the Hantzsch dihydropyridine synthesis under microwave conditions. Synlett 2001:1296–1298. doi:10.1055/s-2001-16043

51. Bandyopadhyay D, Maldonado S, Banik BK (2012) A microwave-assisted bismuth nitrate-catalyzed unique route toward 1,4-dihydropyridines. Molecules 17:2643–2662. doi:10.3390/molecules17032643
52. Kar P, Mishra BG (2013) Silicotungstic acid nanoparticles dispersed in the micropores of Cr-pillared clay as efficient heterogeneous catalyst for the solvent free synthesis of 1,4-dihydropyridines. Chem Eng J 223:647–656. doi:10.1016/j.cej.2013.03.050
53. Bucherer HT, Fischbeck HT (1934) J Prakt Chem 140:69
54. Bergs H (1933) Chem Abstr 27:1001
55. Safari J, Gandomi-Ravandi S, Javadian L (2013) Microwave-promoted facile and rapid synthesis procedure for the efficient synthesis of 5,5-disubstituted hydantoins. Synth Commun 43:3115–3120. doi:10.1080/00397911.2012.730647
56. Handzlik J, Bojarski AJ, Satała G, Kubacka M, Sadek B, Ashoor A, Siwek A, Więcek M, Kucwaj K, Filipek B, Kieć-Kononowicz K (2014) SAR-studies on the importance of aromatic ring topologies in search for selective 5-HT7 receptor ligands among phenylpiperazine hydantoin derivatives. Eur J Med Chem 78:324–339. doi:10.1016/j.ejmech.2014.01.065
57. Strecker A (1854) Ueber einen neuen aus aldehyd—ammoniak und blausäure entstehenden körper. Justus Liebigs Ann Chem 91:349–351. doi:10.1002/jlac.18540910309
58. Prakash GKS, Bychinskaya I, Marinez ER, Mathew T, Olah GA (2013) Nafion–Fe: a new efficient "green" lewis acid catalyst for the ketonic strecker reaction. Catal Lett 143:303–312. doi:10.1007/s10562-012-0958-2
59. Rajabi F, Nourian S, Ghiassian S, Balu AM, Saidi MR, Serrano-Ruiz JC, Luque R (2011) Heterogeneously catalysed Strecker-type reactions using supported Co(ii) catalysts: microwave versus conventional heating. Green Chem 13:3282–3289. doi:10.1039/c1gc15741h
60. Das D, Richers MT, Ma L, Seidel D (2011) The decarboxylative Strecker reaction. Org Lett 13:6584–6587. doi:10.1021/ol202957d
61. Gewald K, Schinke E, Böttcher H (1966) Heterocyclen aus CH-aciden nitrilen, VIII. 2-Amino-thiophene aus methylenaktiven nitrilen, carbonylverbindungen und schwefel. Chem Ber 99:94–100. doi:10.1002/cber.19660990116
62. Revelant G, Dunand S, Hesse S, Kirsch G (2011) Microwave-assisted synthesis of 5-substituted 2-aminothiophenes starting from arylacetaldehydes. Synthesis 2011:2935–2940. doi:10.1055/s-0030-1261032
63. Sridhar M, Rao RM, Baba NHK, Kumbhare RM (2007) Microwave accelerated Gewald reaction: synthesis of 2-aminothiophenes. Tetrahedron Lett 48:3171–3172. doi:10.1016/j.tetlet.2007.03.052
64. Kunda PK, Rao JV, Mukkanti K, Induri M, Reddy GD (2013) Synthesis, anticonvulsant activity and in silco studies of schiff bases of 2-aminothiophenes via guanidine catalyzed Gewald reaction. Trop J Pharm Res 12:566–576. doi:10.4314/tjpr.v12i4.19
65. Zhang H, Yang G, Chen J, Chen Z (2004) Synthesis of thiophene derivatives on soluble polymer-support using Gewald reaction. Synthesis 2004:3055–3059. doi:10.1055/s-2004-834895
66. Fields EK (1952) The synthesis of esters of substituted amino phosphonic acids. J Am Chem Soc 74:1528–1531. doi:10.1021/ja01126a054
67. Kabachnik Martin I, Medved TY (1952) A new method for the synthesis of α-amino phosphoric acids. Dokl Akad Nauk SSSR 83:689
68. Mu X-J, Lei M-Y, Zou J-P, Zhang W (2006) Microwave-assisted solvent-free and catalyst-free Kabachnik-Fields reactions for α-amino phosphonates. Tetrahedron Lett 47:1125–1127. doi:10.1016/j.tetlet.2005.12.027
69. Tibhe GD, Reyes-González MA, Cativiela C, Ordóñez M (2012) Microwave-assisted high diastereoselective synthesis of α-aminophosphonates under solvent and catalyst free-conditions. J Mex Chem Soc 56:183–187
70. Gangireddy CSR, Chinthaparthi RR, Mudumala VNR, Cirandur SR (2014) An elegant microwave assisted one-pot synthesis of di(α-aminophosphonate) pesticides. Arch Pharm 347:819–824. doi:10.1002/ardp.201400213

71. Kindler K (1923) Studien über den mechanismus chemischer reaktionen. Erste abhandlung. Reduktion von amiden und oxydation von aminen. Justus Liebigs Ann Chem 431:187–230. doi:10.1002/jlac.19234310111

72. Agnimonhan FH, Ahoussi L, Kpoviessi SDS, Gbaguidi FA, Kapanda CN, Moudachirou M, Poupaert J, Accrombessi GC (2014) Comparative study of acid and basic catalysis in microwave assistance of Willgerodt-Kindler reaction for thiobenzamides and derivatives synthesis. Int J Biol Chem Sci 8:386–393. doi:10.4314/ijbcs.v8i1.32

73. Zbruyev OI, Stiasni N, Kappe CO (2003) Preparation of thioamide building blocks via microwave-promoted three-component Kindler reactions. J Comb Chem 5:145–148. doi:10. 1021/cc0200538

74. Passerini M, Simone L (1921) Gazz Chim Ital 51:126–129

75. Barreto AFS, Vercillo OE, Andrade CKZ (2011) Microwave-assisted Passerini reactions under solvent-free conditions. J Brazil Chem Soc 22:462–467. doi:10.1590/S0103-50532011000300008

76. Chai C-C, Liu P-Y, Lin C-H, Chen H-C, Wu Y-C, Chang F-R, Pan P-S (2013) Efficient synthesis of boron-containing α-acyloxyamide analogs via microwave irradiation. Molecules 18:9488–9511. doi:10.3390/molecules18089488

77. Ingold M, López GV, Porcal W (2014) Green conditions for Passerini three-component synthesis of tocopherol analogues. ACS Sustainable Chem Eng 2:1093–1097. doi:10.1021/sc5002116

78. Ugi I, Meyr R, Fetzer U, Steinbrückner C (1959) Versuche mit isonitrilen. Angew Chem 71:386. doi:10.1002/ange.19590711110

79. Xing X, Wu J, Feng G, Dai W-M (2006) Microwave-assisted one-pot U-4CR and intramolecular O-alkylation toward heterocyclic scaffolds. Tetrahedron 62:6774–6781. doi:10.1016/j.tet.2006.05.001

80. Tye H, Whittaker M (2004) Use of a Design of Experiments approach for the optimisation of a microwave assisted Ugi reaction. Org Biomol Chem 2:813–815. doi:10.1039/b400298a

81. Jida M, Malaquin S, Deprez-Poulain R, Laconde G, Deprez B (2010) Synthesis of five- and six-membered lactams via solvent-free microwave Ugi reaction. Tetrahedron Lett 51:5109–5111. doi:10.1016/j.tetlet.2010.07.021

82. Jiang B, Wang X, Shi F, Tu S-J, Ai T, Ballew A, Li G (2009) Microwave enabled umpulong mechanism based rapid and efficient four- and six-component domino formations of 2-(2'-azaaryl)imidazoles and anti-1,2-diarylethylbenzamides. J Org Chem 74:9486–9489. doi:10.1021/jo902204s

83. Kulkarni A, Torok B (2010) Microwave-assisted multicomponent domino cyclization-aromatization: an efficient approach for the synthesis of substituted quinolines. Green Chem 12:875–878. doi:10.1039/c001076f

84. Appukkuttan P, Mehta VP, Van der Eycken EV (2010) Microwave-assisted cycloaddition reactions. Chem Soc Rev 39:1467–1477. doi:10.1039/B815717K

85. Bougrin K, Benhida R (2012) Microwave-assisted cycloaddition reactions. In: Microwaves in organic synthesis. Wiley-VCH, Verlag, pp 737–809. doi:10.1002/9783527651313.ch17

86. Ovaska TV, Kyne RE (2008) Intramolecular thermal allenyne [2+2] cycloadditions: facile construction of the 5–6–4 ring core of sterpurene. Tetrahedron Lett 49:376–378. doi:10.1016/j.tetlet.2007.11.042

87. Brummond KM, Chen D (2005) Microwave-Assisted Intramolecular [2+2] Allenic Cycloaddition Reaction for the Rapid Assembly of Bicyclo[4.2.0]octa-1,6-dienes and Bicyclo[5.2.0]nona-1,7-dienes. Org Lett 7:3473–3475. doi:10.1021/ol051115g

88. Hu L, Wang Y, Li B, Du D-M, Xu J (2007) Diastereoselectivity in the Staudinger reaction: a useful probe for investigation of nonthermal microwave effects. Tetrahedron 63:9387–9392. doi:10.1016/j.tet.2007.06.112

89. Uršič U, Grošelj U, Meden A, Svete J, Stanovnik B (2008) Regiospecific [2+2] cycloadditions of electron-poor acetylenes to (Z)-2-acylamino-3-dimethylaminopropenoates: synthesis of highly functionalised buta-1,3-dienes. Tetrahedron Lett 49:3775–3778. doi:10.1016/j.tetlet.2008.04.021

90. Johnstone MD, Lowe AJ, Henderson LC, Pfeffer FM (2010) Rapid synthesis of cyclobutene diesters using a microwave-accelerated ruthenium-catalysed [2+2] cycloaddition. Tetrahedron Lett 51:5889–5891. doi:10.1016/j.tetlet.2010.08.119

91. Himo F, Lovell T, Hilgraf R, Rostovtsev VV, Noodleman L, Sharpless KB, Fokin VV (2005) Copper(I)-catalyzed synthesis of azoles. DFT Study predicts unprecedented reactivity and intermediates. J Am Chem Soc 127:210–216. doi:10.1021/ja0471525

92. Moorhouse AD, Moses JE (2008) Microwave enhancement of a 'one-pot' tandem azidation-'click' cycloaddition of anilines. Synlett 2008:2089–2092. doi:10.1055/s-2008-1078019

93. Taher A, Nandi D, Islam RU, Choudhary M, Mallick K (2015) Microwave assisted azide-alkyne cycloaddition reaction using polymer supported Cu(i) as a catalytic species: a solventless approach. RSC Adv 5:47275–47283. doi:10.1039/C5RA04490A

94. Rasmussen LK, Boren BC, Fokin VV (2007) Ruthenium-catalyzed cycloaddition of aryl azides and alkynes. Org Lett 9:5337–5339. doi:10.1021/ol701912s

95. Louërat F, Bougrin K, Loupy A, Ochoa de Retana AM, Pagalday J, Palacios F (1998) Cycloaddition reactions of azidomethyl phosphonate with acetylenes and enamines. Synthesis of triazoles. Heterocycles 48:161–170. doi:10.3987/COM-97-7997

96. Cantillo D, Gutmann B, Kappe CO (2011) Mechanistic insights on azide–nitrile cycloadditions: on the dialkyltin oxide–trimethylsilyl azide route and a new Vilsmeier–Haack-type organocatalyst. J Am Chem Soc 133:4465–4475. doi:10.1021/ja109700b

97. Díaz-Ortiz Á, de Cózar A, Prieto P, de la Hoz A, Moreno A (2006) Recyclable supported catalysts in microwave-assisted reactions: first Diels-Alder cycloaddition of a triazole ring. Tetrahedron Lett 47:8761–8764. doi:10.1016/j.tetlet.2006.10.006

98. Zheng S, Chowdhury A, Ojima I, Honda T (2013) Microwave-assisted Diels-Alder reactions between Danishefsky's diene and derivatives of ethyl α-(hydroxymethyl)acrylate. Synthetic approach toward a biotinylated anti-inflammatory monocyclic cyanoenone. Tetrahedron 69:2052–2055. doi:10.1016/j.tet.2012.12.079

99. Petronijevic F, Timmons C, Cuzzupe A, Wipf P (2008) A microwave assisted intramolecular-furan-Diels-Alder approach to 4-substituted indoles. Chem Commun 1:106. doi:10.1039/B816989F

100. Kranjc K, Kočevar M (2008) An expedient route to indoles via a cycloaddition/cyclization sequence from (Z)-1-methoxybut-1-en-3-yne and 2H-pyran-2-ones. Tetrahedron 64:45–52. doi:10.1016/j.tet.2007.10.099

101. Gómez VM, Aranda AI, Moreno A, Cossío FP, de Cózar A, Díaz-Ortiz Á, de la Hoz A, Prieto P (2009) Microwave-assisted reactions of nitroheterocycles with dienes. Diels-Alder and tandem hetero Diels–Alder/[3,3] sigmatropic shift. Tetrahedron 65:5328–5336. doi:10. 1016/j.tet.2009.04.065

102. Benedetti E, Kocsis LS, Brummond KM (2012) Synthesis and photophysical properties of a series of cyclopenta[b]naphthalene solvatochromic fluorophores. J Am Chem Soc 134:12418–12421. doi:10.1021/ja3055029

103. Brummond KM, Kocsis LS (2015) Intramolecular didehydro-Diels–Alder reaction and its impact on the structure–function properties of environmentally sensitive fluorophores. Acc Chem Res 48:2320–2329. doi:10.1021/acs.accounts.5b00126

104. Hajbi Y, Suzenet F, Khouili M, Lazar S, Guillaumet G (2007) Polysubstituted 2,3-dihydrofuro[2,3-b]pyridines and 3,4-dihydro-2H-pyrano[2,3-b]pyridines via microwave-activated inverse electron demand Diels-Alder reactions. Tetrahedron 63:8286–8297. doi:10.1016/j.tet.2007.05.112

105. Lee YA, Kim SC (2011) Synthesis of 1,4-dihydropyridine using microwave-assisted aza-Diels–Alder reaction and its application to Amlodipine. J Ind Eng Chem 17:401–403. doi:10.1016/j.jiec.2011.02.031

106. Singh L, Singh Ishar MP, Elango M, Subramanian V, Gupta V, Kanwal P (2008) Synthesis of unsymmetrical substituted 1,4-dihydropyridines through thermal and microwave assisted [4+2] cycloadditions of 1-azadienes and allenic esters. J Org Chem 73:2224–2233. doi:10. 1021/jo702548b

# Chapter 3
# The Use of MW in Organophosphorus Chemistry

György Keglevich, Erika Bálint and Nóra Zs. Kiss

**Abstract** The third chapter summarizes a special field, the application of the microwave (MW) technique in the synthesis of organophosphorus compounds. On the one hand, reactions are shown that are otherwise rather reluctant on traditional thermal heating. On the other hand, reactions are discussed, which, became more efficient (shorter reaction times and higher yields) on MW irradiation. Finally, the simplification of catalytic systems under MW conditions are surveyed.

**Keywords** Microwave · Organophosphorus chemistry · P-heterocycles · Direct esterification · Alkylating esterification · T3P® reagent · C-alkylation · Kabachnik–fields condensation · Arbuzov reaction · Hirao reaction

## 3.1 Introduction

The use of MW technique in general organic syntheses was spread revolutionarily in the last 35 years. As such, this novel approach was, of course, applied also in organophosphorus chemistry from the beginning (~1980), however, the real break-through happened much later. Guenin was the first, who collected the examples of MW-assisted organophosphorus reactions into a review article [1] that was followed by a few others compiled mainly by the author of this chapter [2–7].

G. Keglevich (✉) · N.Zs.Kiss
Department of Organic Chemistry and Technology, Budapest University of Technology and Economics, 1521 Budapest, Hungary
e-mail: gkeglevich@mail.bme.hu

E. Bálint
MTA-BME Research Group for Organic Chemical Technology, 1521 Budapest, Hungary

© Springer International Publishing Switzerland 2016
G. Keglevich (ed.), *Milestones in Microwave Chemistry*,
SpringerBriefs in Green Chemistry for Sustainability,
DOI 10.1007/978-3-319-30632-2_3

## 3.2    Alkylation of Active Methylene Containing Substrates

### 3.2.1    Monoalkylation of P=O-Functionalized CH-Acidic Compounds

It was found that simple active methylene containing compounds underwent C-alkylation by reaction with alkyl halides in the presence of $K_2CO_3$ under solvent-free MW conditions. The message of this discovery is that the phase transfer catalyst can be substituted by MW irradiation [8, 9]. This method was then extended to the alkylation of tetraethyl methylenebisphosphonate (**1a**), diethyl cyanomethylphosphonate (**1b**) and diethyl ethoxycarbonylmethylphosphonate (**1c**) using $K_2CO_3$ or $Cs_2CO_3$ as the base to give the corresponding monoalkylated products (**2a-c**) in variable yields (Scheme 3.1, Table 3.1) [10–12].

The phase transfer catalyzed and MW-assisted alkylations of active methylene containing substrates were summarized [13–15].

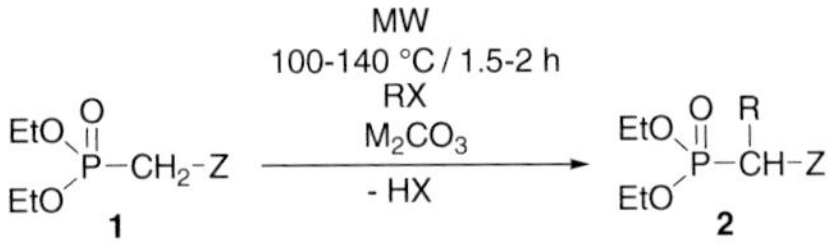

MW
100-140 °C / 1.5-2 h
RX
$M_2CO_3$
- HX

$\underset{EtO}{\overset{EtO}{\phantom{|}}}\overset{O}{\underset{|}{P}}-CH_2-Z \quad \longrightarrow \quad \underset{EtO}{\overset{EtO}{\phantom{|}}}\overset{O}{\underset{|}{P}}\overset{R}{\underset{|}{-}}CH-Z$

**1**                       **2**

Z = P(O)(OEt)₂ (**a**), CN (**b**), CO₂Et (**c**)
RX = EtI, PrBr, BuBr
M = K or Cs

**Scheme 3.1** MW-assisted substitution of active methylene containing compounds

**Table 3.1** Summary of the MW-assisted alkylation of CH-acidic compounds

| Entry | Starting material | RX | $M_2CO_3$ | Solvent | Mode of heating | T/p (°C/bar) | t (h) | Yield of **2** | Ref. |
|---|---|---|---|---|---|---|---|---|---|
| 1 | **1a** | EtI | $Cs_2CO_3$ | – | MW | 140/11 | 1.5 | 80 | [10] |
| 2 | **1a** | ⁿPrBr | $Cs_2CO_3$ | – | MW | 120/6 | 4 | 57[a,b] | [10] |
| 3 | **1b** | ⁿPrBr | $K_2CO_3$ | – | MW | 100/2.5 | 2 | 64 | [11] |
| 4 | **1b** | ⁿBuBr | $K_2CO_3$ | – | MW | 120/3 | 2 | 59 | [11] |
| 5 | **1c** | EtI | $Cs_2CO_3$ | – | MW | 120 | 2 | 70 | [12] |
| 6 | **1c** | ⁿPrBr | $Cs_2CO_3$ | – | MW | 120 | 2 | 71 | [12] |
| 7 | **1c** | ⁿBuBr | $Cs_2CO_3$ | – | MW | 120 | 2 | 70 | [12] |

[a]Proportion in the mixture on the basis of GC
[b]The mixed esters with one or two PrO groups were also present in 33 % and 10 %, respectively

## 3.2.2 Dialkylation of P=O-Functionalized CH-Acidic Compounds

A multi-step variation of the above method led to dialkyl derivatives [16]. Reacting diethyl ethoxycarbonylmethylphosphonate (**1c**) with 1.2 equivalents of alkyl iodides in the presence of 1 equivalent of $Cs_2CO_3$ at 120 °C for 2 h, and repeating the treatment of the crude product using 2 equivalents of the same alkylating agent in the presence of 1.5 equivalents of $Cs_2CO_3$ for four times, the dialkylated products (**3**) were obtained in 41–64 % yields (Scheme 3.2/**A**). In a similar way, applying propyl iodide, butyl bromide, benzyl bromide in the first step, and ethyl iodide in the second, third and fourth steps, the mixed alkyl derivatives **4** were prepared in yields of 36–40 % (Scheme 3.2/**B**).

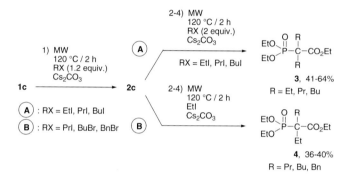

**Scheme 3.2**  Dialkylation of diethyl ethoxycarbonylmethylphosphonate

## 3.3 Esterification-Related Reactions

### 3.3.1 Direct Esterification of Phosphinic Acids

It is well-known that phosphinic acids (**5**) generally do not undergo direct esterification with alcohols to afford phosphinates (**6**) (Scheme 3.3/**A**). There are only a few examples for the direct esterification of phosphinic acids [17]. For this, the esters of phosphinic acids (**6**) are, in most cases, synthesized by the reaction of phosphinic chlorides (**7**) with alcohols in the presence of a base (Scheme 3.3/**B**) [18–20].

**Scheme 3.3**  Synthetic routes to alkyl phosphinates

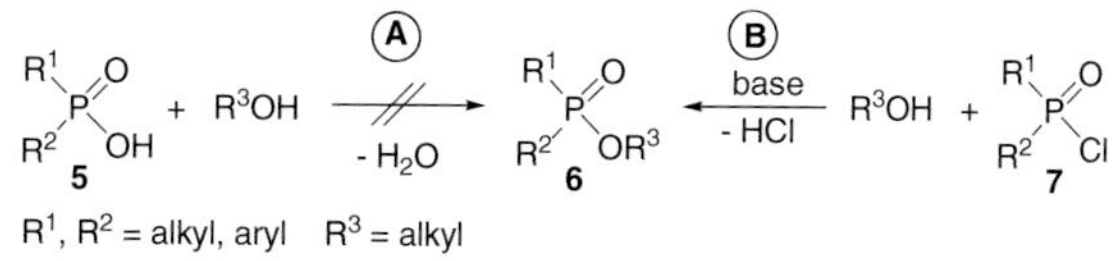

The generally applied esterification method (Scheme 3.3/**B**) has the disadvantage of requiring the use of relatively expensive *P*-chlorides (**7**). Moreover, the hydrogen chloride formed as the by-product should be removed by a base, and this method is not atomic efficient.

We have recently found that a series of phosphinic acids underwent direct esterification with longer chain alcohols on MW irradiation at around 170–220 °C [21]. The first example was the esterification of phenyl-*H*-phosphinic acid (**8**) at 170 °C to give the corresponding phosphinates (**9**) in yields of 73–90 % (Scheme 3.4).

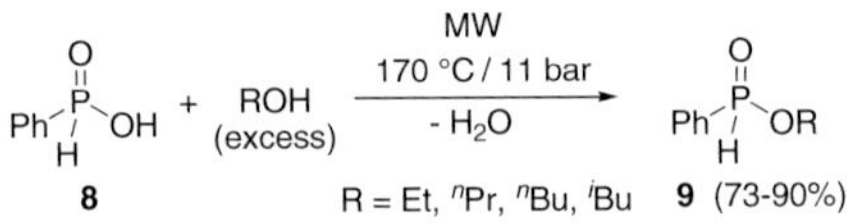

**Scheme 3.4** Direct esterification of phenyl-*H*-phosphinic acid under MW conditions

This esterification was assumed to take place via the tervalent tautomer of the phenyl-*H*-phosphine acid, but no evidence was presented [22].

The MW-assisted esterification of cyclic phosphinic acids, such as 1-hydroxy-3-phospholene oxides (**10**), 1-hydroxy-phospholane oxides (**11**) and 1-hydroxy-1,2,3,4,5,6-hexahydrophosphinine oxides (**12**) was carried out at 180–235 °C in the presence of *ca.* 15-fold excess of the alcohols to afford the corresponding alkyl phosphinates (**13–15**) in variable yields (Schemes 3.5, 3.6 and 3.7, Table 3.2) [23–26].

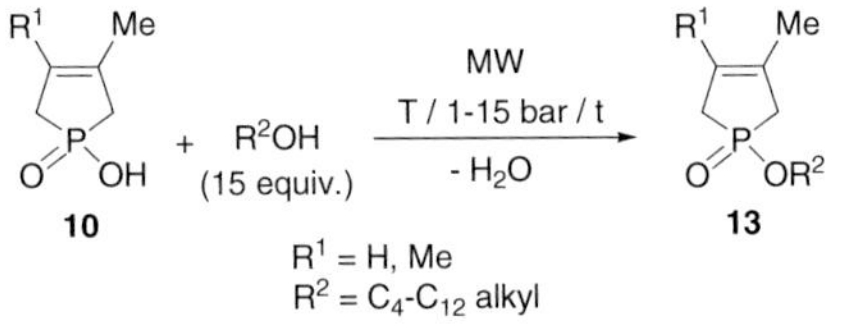

**Scheme 3.5** MW-assisted direct esterification of 1-hydroxy-3-phospholene 1-oxides

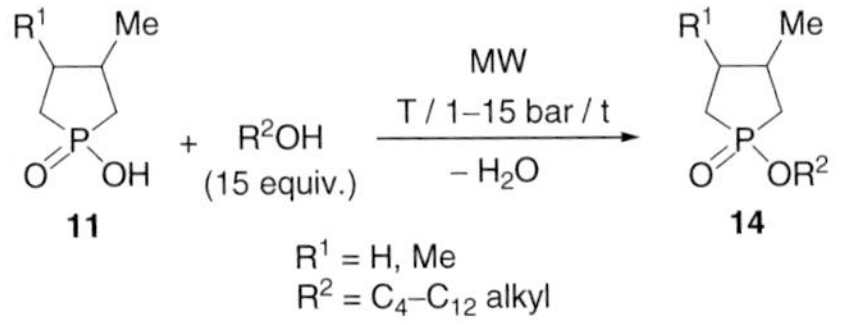

**Scheme 3.6** MW-assisted direct esterification of 1-hydroxyphospholane 1-oxides

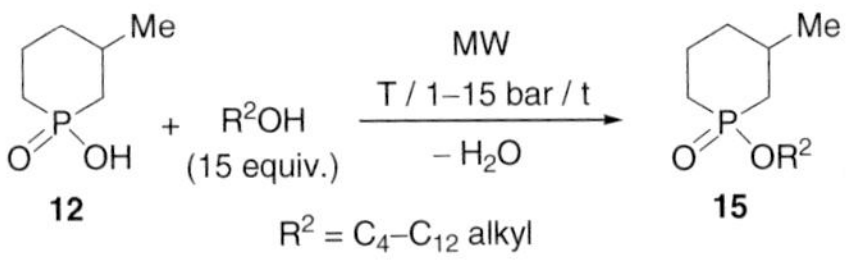

**Scheme 3.7** MW-assisted direct esterification of a 1-hydroxy-1,2,3,4,5,6-hexahydrophosphinine 1-oxide

**Table 3.2** MW-assisted direct esterification of cyclic phosphinic acids (**10–12**)

| Phosphinic acid | $R^2$ | T (°C) | t (h) | Isomeric composition | Yield of **13–15** MW (%) | Entry |
|---|---|---|---|---|---|---|
| Me (**10**, $R^1$ = H) | $^n$Bu | 200 | 2 | – | 58 (11) | 1 |
| | $^n$Pent | 220 | 2.5 | – | 94 | 2 |
| | $^i$Pent | 235 | 3 | – | 74 | 3 |
| | $^n$Oct | 220 | 2 | – | 71 | 4 |
| | $^i$Oct | 220 | 2 | – | 76[a] | 5 |
| | $^n$Dodec | 230 | 2 | – | 95 | 6 |
| Me  Me (**10**, $R^1$ = Me) | $^n$Pr | 180 | 4 | – | 20 | 7 |
| | $^n$Bu | 220 | 3 | – | 60[b] | 8 |
| | $^i$Bu | 200 | 2 | – | 30 | 9 |
| | $^n$Pent | 235 | 3 | – | 67 | 10 |
| | $^i$Pent | 235 | 4 | – | 57 | 11 |
| | $^n$Oct | 230 | 2 | – | 95 | 12 |
| | $^i$Oct | 220 | 2.5 | – | 82[c] | 13 |
| | $^n$Dodec | 230 | 2 | – | 95 | 14 |
| Me (**11**, $R^1$ = H) | $^n$Bu | 230 | 3 | ~ 50–50 | 45 | 15 |
| | $^n$Pent | 235 | 3 | ~ 50–50 | 79 | 16 |
| | $^i$Pent | 235 | 4 | ~ 50–50 | 59 | 17 |
| | $^n$Oct | 230 | 4 | ~ 50–50 | 74 | 18 |
| | $^i$Oct | 220 | 3 | ~ 50–50 | 86 | 19 |
| Me  Me (**11**, $R^1$ = Me) | $^n$Bu | 210 | 3 | ~ 60–20–20 | 54 | 20 |
| | $^n$Pent | 235 | 5 | ~ 70–15–15 | 60 | 21 |
| | $^i$Pent | 235 | 6 | ~ 64–19–17 | 56 | 22 |
| | $^n$Oct | 230 | 4 | ~ 60–20–20 | 70 | 23 |
| | $^i$Oct | 220 | 4 | ~ 66–19–15 | 50 | 24 |
| Me (**12**) | $^n$Bu | 230 | 3 | ~ 69–31 | 45 | 25 |
| | $^n$Oct | 235 | 4 | ~ 66–34 | 62 | 26 |
| | $^i$Oct | 235 | 6 | ~ 69–31 | 54 | 27 |

[a]In the thermal variation, the yield was 22 %
[b]In the thermal variation, the yield was 13 %
[c]In the thermal variation, the yield was 24 %

The method developed by us seems to be of more general value. It was also found that the esterification of phosphinic acids is thermoneutral and controlled kinetically. The reaction enthalpies were found to fall in the range of 0.1–4.0 kJ mol$^{-1}$, while the activation enthalpies were in the range of 102.0–161.0 kJ mol$^{-1}$ [25, 27]. Interestingly, the analogous thioesterifications were reluctant and incomplete under MW conditions [28].

## 3.3.2   Esterification in the Presence of the T3P® Reagent

The phenyl-*H*-phosphinic acid (**8**) could be esterified with simple alcohols efficiently and under mild conditions in the presence of 1.1 equivalents of the T3P®
reagent as the activating agent (Scheme 3.8) [29].

**Scheme 3.8** T3P®-mediated esterification of phenyl-*H*-phosphinic acid

$$\begin{array}{cc} & \text{1) 25 °C / 0.5–1 h} \\ & \text{T3P® (1.1 equiv.)} \\ & \text{EtOAc} \\ \mathbf{8} \; \xrightarrow{\hspace{3cm}} \; \mathbf{9} \\ & \text{2) ROH (3 equiv.)} \quad \text{81–95\%} \end{array}$$

R = Me, Et, $^n$Pr, $^i$Pr, Bu, $^i$Bu, $^s$Bu, $^t$Bu

It was found that the use of only 0.66 equivalents of the T3P® reagent was also
enough to reach yields of 78–91 % [29].

1-Hydroxy-3-phospholene 1-oxides (**10**) also underwent a similar T3P®-pro-
moted direct esterification (Scheme 3.9) [29, 30].

**Scheme 3.9** T3P®-mediated esterification of 1-hydroxy-3-phospholene 1-oxides

$$\begin{array}{cc} & \text{1) 25 °C / 0.5–3 h} \\ & \text{T3P® (1.1 equiv.)} \\ & \text{EtOAc} \\ \mathbf{10} \; \xrightarrow{\hspace{3cm}} \; \mathbf{13} \\ & \text{2) ROH (3 equiv.)} \quad \text{61–95\%} \end{array}$$

R = Me, Et, $^n$Pr, $^i$Pr, Bu, $^i$Bu, $^s$Bu, Pent, $^i$Pent, 3-pentyl, $^c$Hex, Bn

However, the quantity of the T3P® reagent could be decreased to 0.66 equivalents only under MW conditions and working at 85 °C [29].

1-Hydroxyphospholane 1-oxides (**11**) were also converted to the corresponding
phosphinates (**14**) in a similar way (Scheme 3.10) [29].

**Scheme 3.10** T3P®-mediated esterification of 1-hydroxyphospholane 1-oxides

$$\begin{array}{cc} & \text{1) 25 °C / 1–2 h} \\ & \text{T3P® (1.1 equiv.)} \\ & \text{EtOAc} \\ \mathbf{11} \; \xrightarrow{\hspace{3cm}} \; \mathbf{14} \\ & \text{2) ROH (3 equiv.)} \quad \text{54–81\%} \end{array}$$

R = Me, Et, $^n$Pr, $^i$Pr, $^n$Bu, $^i$Bu, $^s$Bu

The use of only 0.66 equivalents of the T3P® reagent at 85 °C under MW
conditions furnished the phosphinates (**14**) in somewhat lower (55–67 %) yields
[29].

The role of the T3P® reagent is to activate the phosphinic acids (**8**, **10** and **11**) by
converting them to the corresponding mixed anhydrides represented by general
structure **16** (Scheme 3.11).

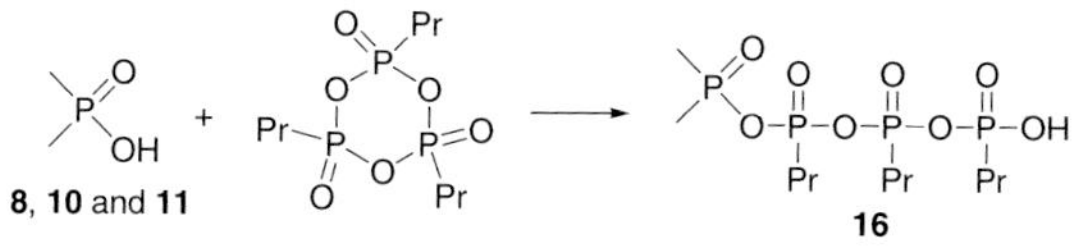

**Scheme 3.11** Activation of phosphinic acids

8, 10 and 11                                                                                     16

## 3.3.3 Alkylating Esterification of Phosphinic Acids

Phosphinic esters may also be synthesized by alkylating esterification utilizing the MW and phase transfer catalytic (PTC) techniques. This is shown on the example of the alkylation of 1-hydroxy-3-phospholene oxides **10** (Scheme 3.12, Table 3.3) [24, 31]. During the alkylations, similarly to that of phenols, the combined application of MW irradiation and PTC was found to be synergistic [32, 33].

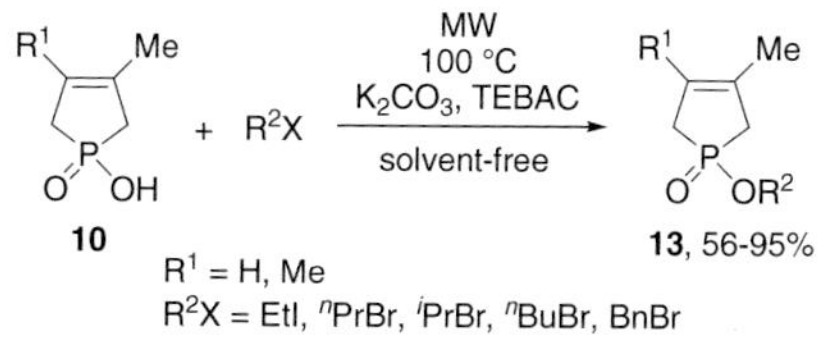

**Scheme 3.12** Alkylating esterification of 1-hydroxy-3-phospholene 1-oxides

$R^1$ = H, Me
$R^2X$ = EtI, $^n$PrBr, $^i$PrBr, $^n$BuBr, BnBr

**Table 3.3** Alkylating esterification of phospholene oxides under solvent-free MW conditions

| $R^1$ | $R^2X$ | TEBAC (%) | t (h) | Yield of **13** (%) |
|-------|--------|-----------|-------|---------------------|
| H | EtI | – | 1 | 80 |
| H | EtI | 5 | 1 | 90 |
| H | $^n$PrBr | – | 1 | 73 |
| H | $^n$PrBr | 5 | 1 | 94 |
| H | $^i$PrBr | – | 1.5 | 42 |
| H | $^i$PrBr | 5 | 1 | 65 |
| Me | EtI | – | 1 | 83 |
| Me | EtI | 5 | 1 | 95 |
| Me | $^n$PrBr | – | 1 | 49 |
| Me | $^n$PrBr | 5 | 1 | 90 |
| Me | $^i$PrBr | – | 1.5 | 18 |
| Me | $^i$PrBr | 5 | 1 | 56 |

The O-alkylations could, of course, be extended to 1-hydroxyphospholane oxides and to a 1-hydroxy-1,2,3,4,5,6-hexahydrophosphinine oxide [24].

The alkylation of thermally unstable cyclic phosphinic acids, such as a 1-hydroxy-3-phosphabicyclo[3.1.0]hexane 3-oxide and 1-hydroxy-1,2-dihydrophosphinine oxide was performed using $K_2CO_3$ in acetone at 63 °C [34].

### 3.3.4   Transesterification Reactions of Dialkyl Phosphites (H-Phosphonates)

It was found that dialkyl phosphites (**17**) underwent alcoholysis on exposing their alcoholic solutions to MW irradiation above 100 °C. In most cases, the three possible phosphites (**18**, **19** and **20**) were present in the mixture (Scheme 3.13, Table 3.4). The reaction could be controlled to give the dialkyl phosphite with two different alkyl groups (**19**) as the major component, or to provide the fully transesterified product (**20**) as an almost exclusive product. Comparative thermal experiments were also performed [35].

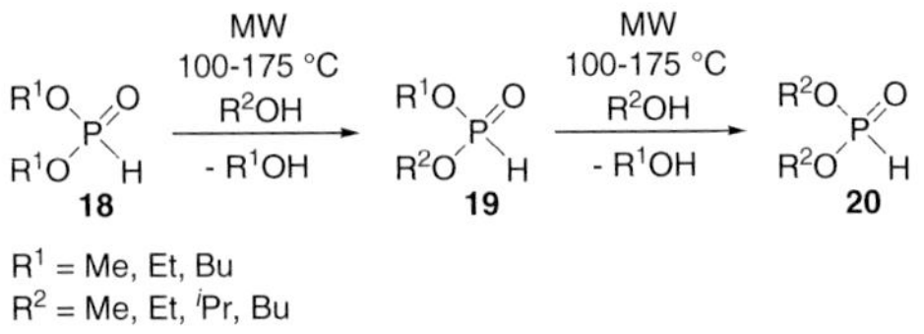

$R^1$ = Me, Et, Bu
$R^2$ = Me, Et, $^i$Pr, Bu

**Scheme 3.13**   Transesterification of dialkyl phosphites

**Table 3.4**   Results of the reactions of dialkyl phosphites with alcohols

| $R^1$ | $R^2$ | Equivalents of $R^2$OH | T (°C) | t (min) | p (bar) | Composition (%) | | |
|-------|-------|------------------------|--------|---------|---------|------|------|------|
| | | | | | | **18** | **19** | **20** |
| Me | Et | 5 | 100 | 120 | 3 | 29 | **56** | 15 |
| Me | Et | 50 | 125 | 60 | 6 | 2 | 37 | 61 |
| Me | Et | 50 | 175 | 40 | 15 | 0 | 4 | **96** |
| Et | Me | 50 | 125 | 60 | 8 | 48 | **40** | 12 |
| Et | Me | 50 | 150 | 60 | 15 | 1 | 19 | **80** |
| Et | Me | 50 | 175 | 40 | 19 | 0 | 21 | 79 |
| Me | Bu | 25 | 125 | 60 | 3 | 0 | **40** | 60 |
| Me | Bu | 50 | 150 | 90 | 6 | 0 | 0 | **100** |
| Et | Bu | 25 | 125 | 60 | 3 | 25 | **54** | 21 |
| Et | Bu | 50 | 175 | 40 | 9 | 0 | 2 | **98** |

As can be seen from Table 3.4, the outcome of alcoholysis depended on the nature of the reagents, the molar ratio of alcohol and substrate, the temperature and reaction time. Using the alcohol in a less excess and applying lower temperatures, the proportion of the mixed dialkyl phosphites (**19**) was 40–56 %. Increasing the proportion of the alcohols and elevating the temperature, the fully transesterified *H*-phosphonate (**20**) became the predominating component. The dialkyl phosphites with mixed alkyl groups are valuable building blocks.

The alcoholysis of diethyl phosphite with ethylene glycol may principally lead to mixed ester **21**, fully transesterified product **22** and bis($H$-phosphonate) **23** (Scheme 3.14, Table 3.5). The composition of the mixture depended on the ratio of the starting materials and the temperature [36].

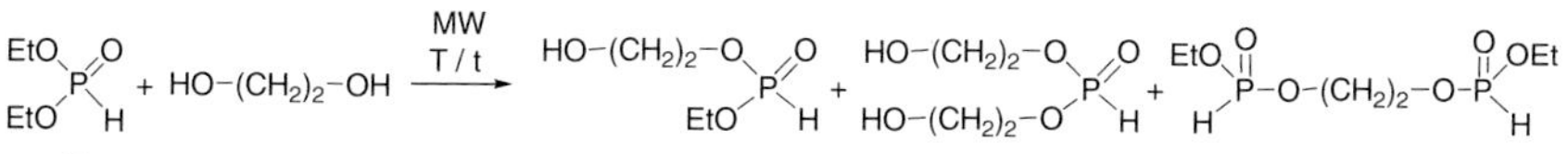

**Scheme 3.14** MW-assisted alcoholysis of diethyl phosphite with ethylene glycol

**Table 3.5** MW-assisted alcoholysis of diethyl phosphite with ethylene glycol under different conditions

| Entry | Molar ratio | | T (°C) | t (h) | Conversion (%) | Product composition (%) | | |
|---|---|---|---|---|---|---|---|---|
| | $(EtO)_2P(O)H$ | $(HOCH_2)_2$ | | | | **21** | **22** | **23** |
| 1 | 1 | 1 | 120 | 6 | 55 | 76 | 13 | 11 |
| 2 | 1 | 1 | 140 | 1[a] | 67 | **79** | 11 | 9 |
| 3 | 1 | 4 | 140 | 1 | 64 | 59 | 41 | 0 |
| 4 | 1 | 8 | 140 | 1 | 39 | 41 | **59** | 0 |
| 5 | 2 | 1 | 140 | 3 | 59 | 75 | 0 | **25** |

[a]On a prolonged heating (3 h), a considerable amount of by-products were formed

The maximum proportion (79 %) of the mixed ester (**21**) was obtained at 140 °C using the reactants in a 1:1 ratio. The bis(hydroxyethyl) derivative (**22**) was formed in a relative proportion of 59 %, if the glycol was measured in an eight-fold quantity. The bis(phosphono) derivative (**23**) was present in 25 % at a $(EtO)_2P(O)H–(HOCH_2)_2$ molar ratio of 2:1.

In the MW-assisted reaction with diethyl phosphite, ethanolamine acted as an O–nucleophile. However, when ethanolamine was applied in only a 1–2 equivalents quantity, not alcoholysis, but mono- and diethylation of the ethanolamine took place. Using ethanolamine in a 4–10-fold quantity at 140 °C for 20 min, different ratios of the mixed phosphinate (**24**) and the fully transesterified product (**25**) were obtained (Scheme 3.15, Table 3.6) [36].

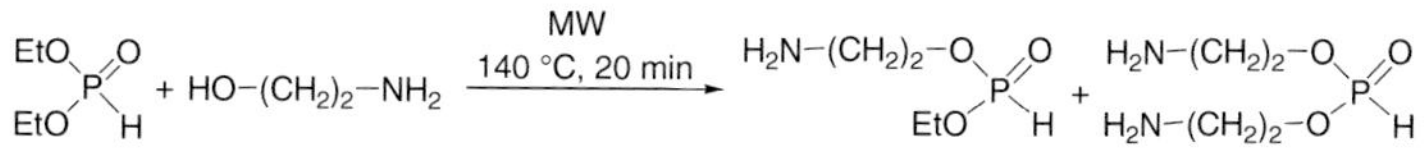

**Scheme 3.15** MW-assisted alcoholysis of diethyl phosphite with ethanolamine

**Table 3.6** MW-assisted alcoholysis of diethyl phosphite with ethanolamine at 140 °C for 20 min

| Entry | Molar ratio | | Conversion (%) | Product composition (%) | |
|-------|-------------|--------------|----------------|----|----|
| | $(EtO)_2P(O)H$ | $HO(CH_2)_2NH_2$ | | **24** | **25** |
| 1 | 1 | 4 | 100 | 45 | 55 |
| 2 | 1 | 8 | 100 | 22 | 78 |
| 3 | 1 | 10 | 100 | 15 | **85** |

From the point of view of bis(aminoethyl)phosphite (**25**), the best experiment was, when ethanolamine was measured in a 10 equivalents quantity and the reaction was performed at 140 °C for 20 min.

## 3.4   Amidation of Phosphinic Acids

Phosphinic acids fail to undergo reaction with amines on heating. The usual synthesis of phosphinic amides (**27**) involves the reaction of phosphinic chlorides (**7**) with amines (Scheme 3.16). The direct amidations of phosphinic acids **5** attempted under MW irradiation remained incomplete with conversions of 30–33 % (Scheme 3.16) [37]. The ammonium salts of type **26** are intermediates in these amidations. The unreactivity even on MW is the consequence of the endothermicity of the direct amidations ($\Delta H^0 = 17.1$–$39.0$ kJ mol$^{-1}$). It can be concluded that the amidation of phosphinic acids is controlled thermodynamically [25].

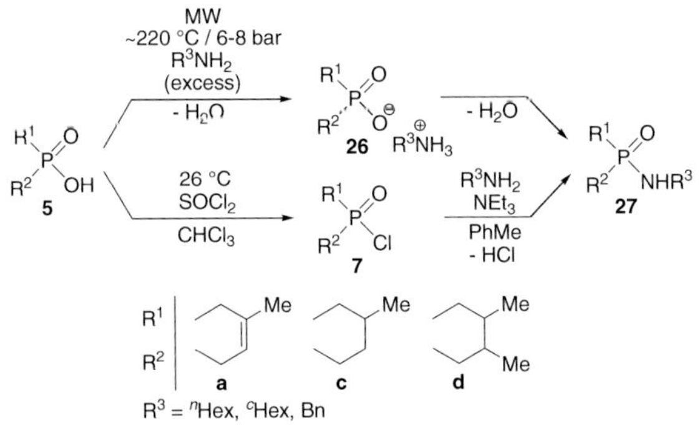

**Scheme 3.16** Possible ways for the amidation of phosphinic acids

In the light of this fact, it is better to carry out the amidations under discussion via the traditional way, using phosphinic chlorides (**7**) as intermediates. However, it was observed that in the reaction of 1-chloro-3-methyl-3-phospholene oxide (**7a**)

with primary amines, bis(phosphinoyl)amine type side products (**28**) were also formed [38].

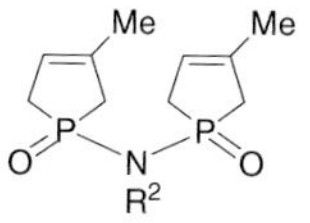

$R^2 = {}^n$Hex, $^c$Hex, Bu

**28**

Moreover, fine-tuning the molar ratio of the components and the addition technique, it was possible to obtain the bis(phosphinoyl)amine (**28**) as the exclusive product.

It was also possible to utilize this protocol in the synthesis of mixed derivatives represented by structure **29** [39].

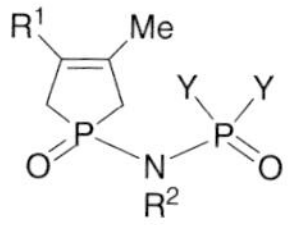

$R^1 = $ H, Me
$R^2 = $ Bu, Bn
$Y = $ Ph, EtO, PhO

**29**

## 3.5   P–C Coupling Reactions

The Hirao-reaction comprising the P–C coupling of a vinyl halide/aryl halide/hetaryl halide and a >P(O)H species has become an important tool to synthesize phosphonates, phosphinates and phosphine oxides [40, 41]. This reaction model prompted many chemists to elaborate "green" variations.

It was interesting to find that there is no need to use the expensive $Pd(Ph_3P)_4$ catalyst in the coupling reaction of dialkyl phosphites with bromobenzene, as $Pd(OAc)_2$ also catalyses the Hirao reaction in the absence of any added P-ligand. A MW-assisted solvent-free accomplishment was the best choice [42, 43]. The arylphosphonates (**30**) were obtained in 69–93 % yields (Scheme 3.17).

**Scheme 3.17** Pd-catalyzed coupling reaction of dialkyl phosphites and aryl bromides

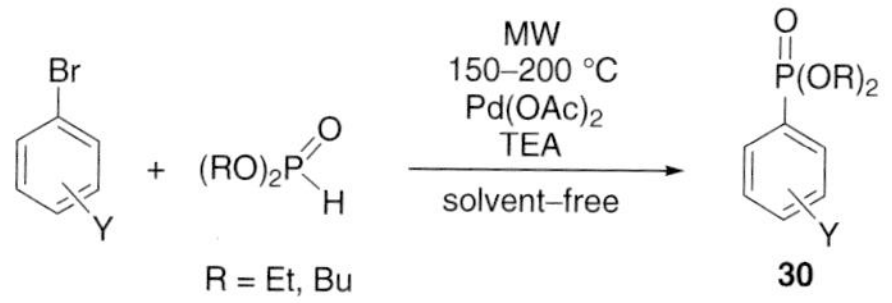

Y = H, 4-MeO, 3-MeO, 4-$^t$Bu, 4-Pr, 4-Et, 4-Me, 3-Me, 4-F, 3-F, 4-Cl, 3-Cl, 4-CO$_2$Et, 3-CO$_2$Et, 4-COMe, 3-COMe

The use of alkyl phenyl-*H*-phosphinates in reaction with bromobenzene led to alkyl diphenylphosphinates (**31**) (Scheme 3.18).

**Scheme 3.18** Pd-catalyzed coupling reaction of alkyl phenyl-*H*-phosphinates and bromobenzene

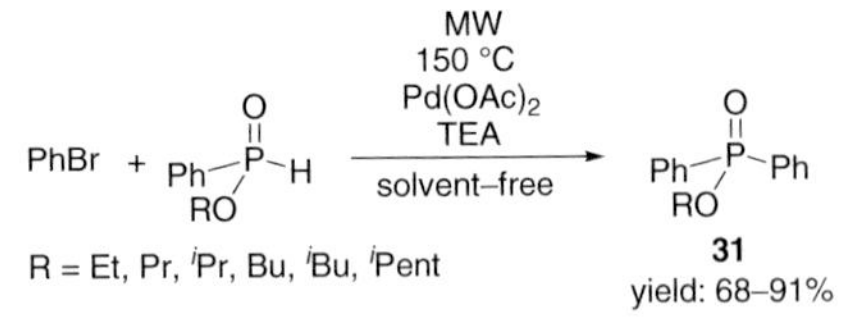

Dibenzo[*c.e*][1,2]oxaphosphorine oxide was also utilized as a P-reagent.

Extending the reaction to secondary phosphine oxides, the products of the P–C coupling reaction are phosphine oxides (**32**) (Scheme 3.19).

**Scheme 3.19** Pd-catalyzed coupling reaction of secondary phosphine oxides and aryl bromides

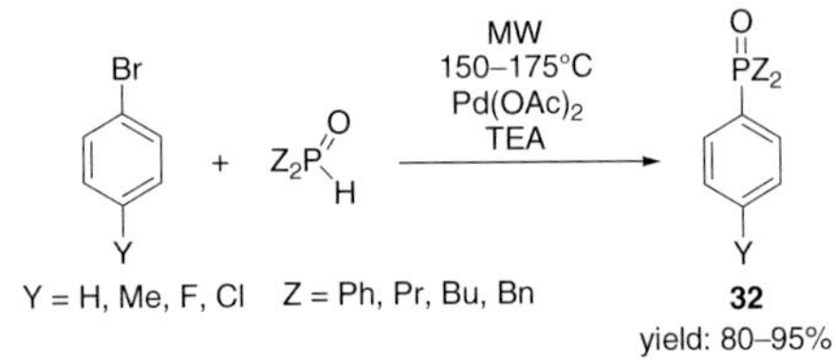

A P-ligand-free Pd-catalyzed method was also utilized in the reaction of 2-nitro-5-bromoanisole with diethyl phosphite using K$_2$CO$_3$ as the base and xylene as the solvent [44].

It was interesting to find that the MW-assisted P-ligand-free accomplishment worked also with NiCl$_2$ as the catalyst. In these variations, acetonitrile had to be used due to the heterogeneity of the reaction mixtures. The series of arylphosphonates (**30**), alkyldiphenylphosphinates (**31**) and diaryl-phenylphosphine oxides (**33**) were prepared as shown in Schemes 3.20, 3.21 and 3.22 [45].

**Scheme 3.20** Ni-catalyzed coupling reaction of alkyl phenyl-*H*-phosphinates and aryl bromides

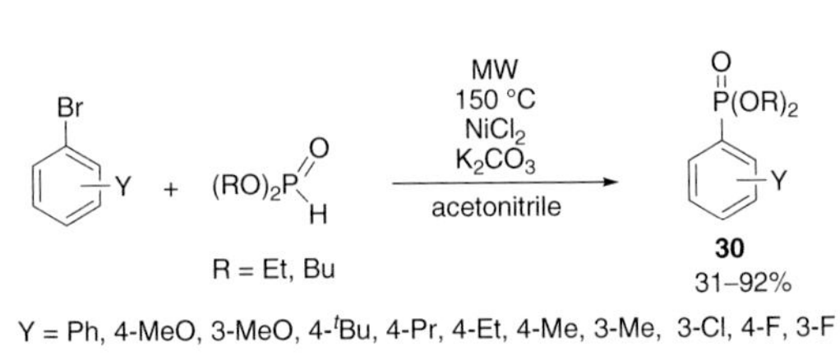

**Scheme 3.21** Ni-catalyzed coupling reaction of dialkyl phosphites and bromobenzene

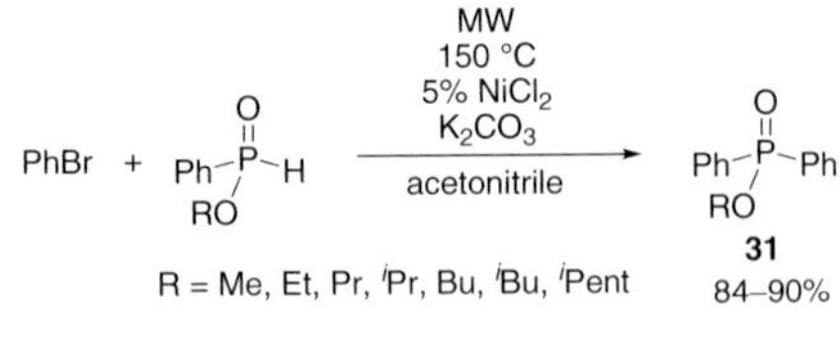

**Scheme 3.22** Ni-catalyzed coupling reaction of secondary phosphine oxides and bromobenzene

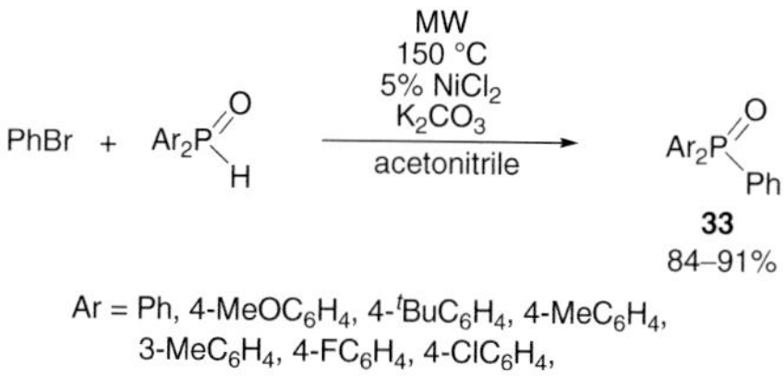

$$PhBr + Ar_2P\overset{O}{\underset{H}{\diagdown}} \xrightarrow[\text{acetonitrile}]{\begin{array}{c}MW \\ 150\ ^\circ C \\ 5\%\ NiCl_2 \\ K_2CO_3\end{array}} Ar_2P\overset{O}{\underset{Ph}{\diagdown}}$$

**33**
84–91%

Ar = Ph, 4-MeOC$_6$H$_4$, 4-$^t$BuC$_6$H$_4$, 4-MeC$_6$H$_4$, 3-MeC$_6$H$_4$, 4-FC$_6$H$_4$, 4-ClC$_6$H$_4$,

The P-ligand-free approaches are environmentally-friendly as save costs and environmental burdens. Other MW-assisted variations of the Hirao-reaction were also described, but they applied P-ligands [46–49].

The P–C coupling of halobenzoic acids and diarylphosphine oxides was performed in the absence of any catalyst in the presence of $K_2CO_3$ in water under MW conditions (Scheme 3.23) [50].

**Scheme 3.23** Catalyst-free coupling reaction of diarylphosphine oxides and halobenzoic acids

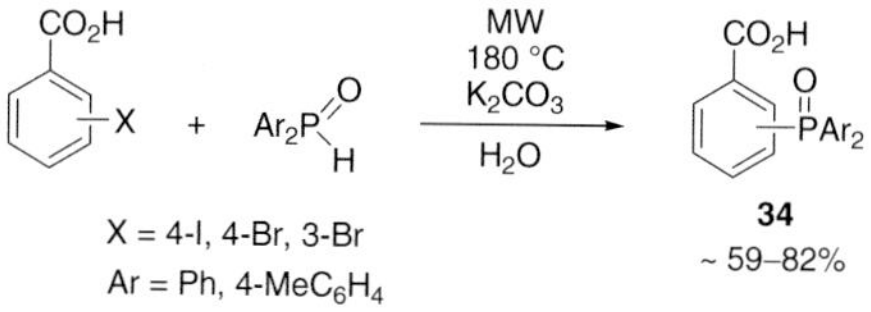

X = 4-I, 4-Br, 3-Br
Ar = Ph, 4-MeC$_6$H$_4$

**34**
~ 59–82%

This protocol was then utilized for the synthesis of a mixed triarylphosphine oxide (**35**) (Scheme 3.24).

**Scheme 3.24** Synthesis of trialkyl-substituted phosphine oxide with different aryl groups

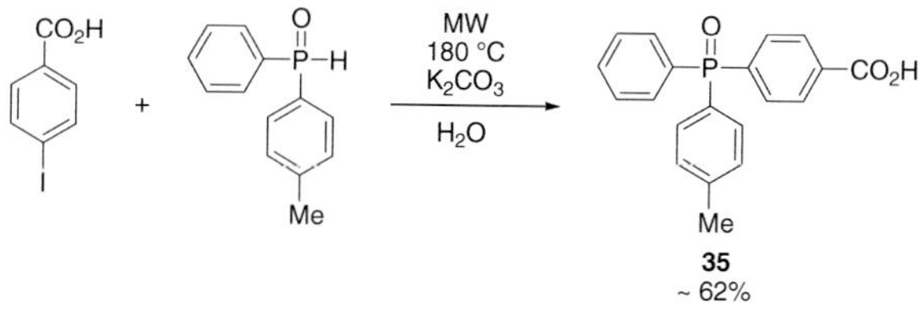

**35**
~ 62%

Till data, the P-ligand-free variation may be the most attractive protocol for the Hirao reaction. As a matter of fact, in the P-ligand-free cases, the trivalent tautomer form of the >P(O)H reagent may act as the ligand. The "green" accomplishments have been summarized [51, 52].

## 3.6    Arbuzov Reactions

Beside the Hirao-reaction, the Arbuzov-reaction is also a suitable method for the preparation of arylphosphonates [53]. However, special protocols are necessary to overcome the decreased reactivity of the aryl halides.

Arylphosphonates (**36**) are the products of the MW-assisted catalytic Arbuzov reaction of aryl bromides with triethyl phosphite. In the presence of a Ni salt catalyst, the phosphonates (**36**) could be prepared in yields of 67–86 % (Scheme 3.25) [54].

**Scheme 3.25** MW-assisted Arbuzov reaction of aryl halides and triethyl phosphite

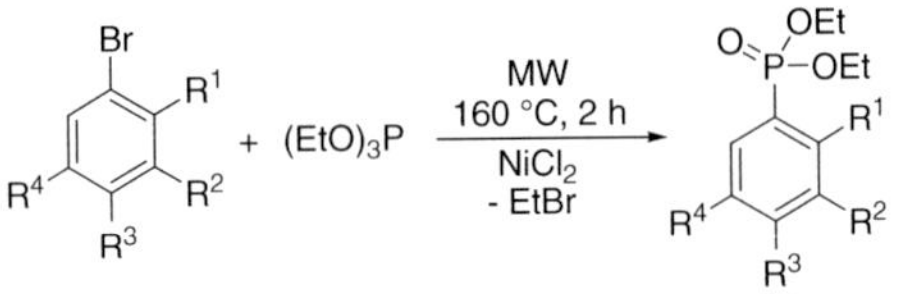

| | $R^1$ | $R^2$ | $R^3$ | $R^4$ |
|---|---|---|---|---|
| **a** | H | H | H | H |
| **b** | $CH_3$ | H | H | H |
| **c** | H | H | $CH_3$ | H |
| **d** | H | H | cyclohexyl | H |
| **e** | H | H | $OCF_3$ | H |
| **f** | H | H | $CF_3$ | H |
| **g** | H | H | $OCH_3$ | H |
| **h** | H | H | $C(O)CH_3$ | H |
| **i** | F | H | H | H |
| **j** | H | F | H | H |
| **k** | H | H | F | H |
| **l** | F | H | H | F |

## 3.7    Phospha-Michael Additions

Simple phospha-Michael reactions, such as the addition of dialkyl phosphites or diphenylphosphine oxide to methyl vinyl ketone or cyclohexene-2-one were performed using NaOR/ROH (R = Me, Et), NaOH/$H_2O$ under PTC or 1,8-diazabicycloundec-7-ene (DBU). There was no need to apply MW irradiation [55]. Phospha-Michael reactions of methyl vinyl ketone with *P*-heterocyclic nucleophiles deriving from a dibenzo-1,2-oxaphosphorine oxide or 1,3,2-dioxaphosphorine oxide were carried out in the presence of DBU [56]. However, the addition of dialkyl phosphites, dibenzo-1,2-oxaphosphorine oxide and diphenylphosphine oxide to the electron-poor double-bond of a 1,2-dihydrophosphinine oxide required a greater activation, so that first the >P(O)H species had to be converted to the corresponding anion by deprotonation with trimethylaluminum [57, 58]. In these reactions, MW irradiation was unable to enhance the addition of the >P(O)H species to the not too reactive CH=CH–P(O)< unsaturation of the reactants.

However, MW irradiation was useful in the addition of dialkyl phosphites and diphenylphosphine oxide to the double-bond of 1-phenyl-2-phospholene 1-oxide (37) (Scheme 3.26) [59]. In these cases, the adducts (38) were formed as 1:1 mixtures of two isomers.

**Scheme 3.26** Michael addition of >P(O)H species to 1-phenyl-2-phospholene 1-oxide

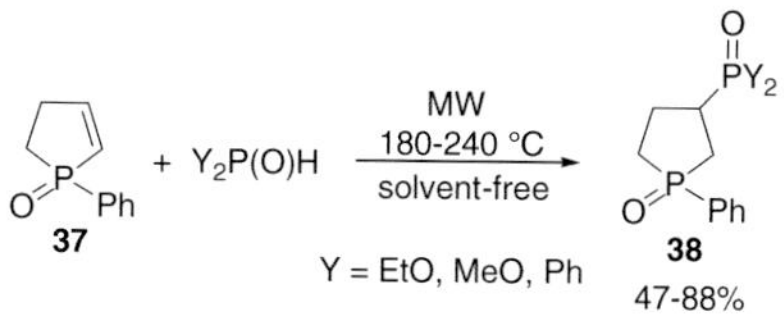

Using dialkyl phosphites pre-reacted with trimethylaluminum, the above additions took place more efficiently (in 89–93 % yields), and were selective leading to only one isomer.

MW addition promoted the addition of dialkyl phosphites, ethyl phenyl-*H*-phosphinate and diphenylphosphine oxide to the reactive unsaturation of *N*-phenyl and *N*-methylmaleimide, as well as maleic anhydride (Schemes 3.27 and 3.28) [60]. In most cases, the reactions were performed in the absence of any solvent. Products **39** and **40** were obtained, with one exception, in good yields.

**Scheme 3.27** Michael addition of >P(O)H species to maleimide derivatives

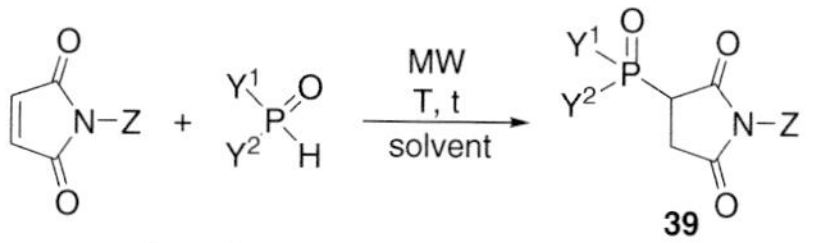

| Z | Y$^1$ | Y$^2$ | T (°C) | t (h) | solvent | yield of **39** (%) |
|---|---|---|---|---|---|---|
| Ph | EtO | EtO | 175 | 2.5 | - | 83 |
| Ph | MeO | MeO | 175 | 2.5 | - | 81 |
| Me | EtO | EtO | 175 | 3 | - | 95 |
| Me | MeO | MeO | 175 | 3 | - | 71 |
| Ph | Ph | EtO | 175 | 2.5 | - | 92 |
| Me | Ph | EtO | 175 | 3 | - | 80 |
| Ph | Ph | Ph | 120 | 3 | MeCN | 97 |
| Me | Ph | Ph | 120 | 3 | MeCN | 98 |

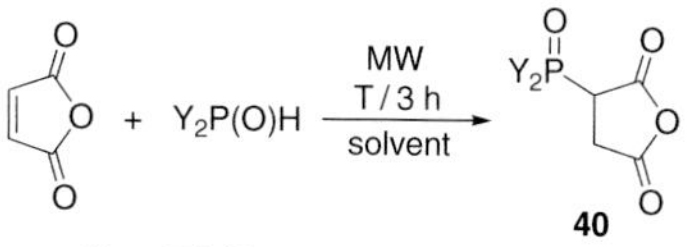

| Y | T (°C) | solvent | yield of **40** (%) |
|---|---|---|---|
| EtO | 120 | - | 49 |
| Ph | 175 | MeCN | 66 |

**Scheme 3.28** Michael addition of > P(O)H species to maleic anhydride

Depending on the molar ratio of the reactants and the conditions (temperature and reaction time), the addition of dialkyl phosphites or diphenylphosphine oxide to the triple bond of dimethyl acetylenedicarboxylate resulted in the formation of a comparable mixture of the corresponding monoadduct (**41**) and bisadduct (**42**), or the bisadduct (**42**) as the predominating or exclusive product (Scheme 3.29, Table 3.7) [61].

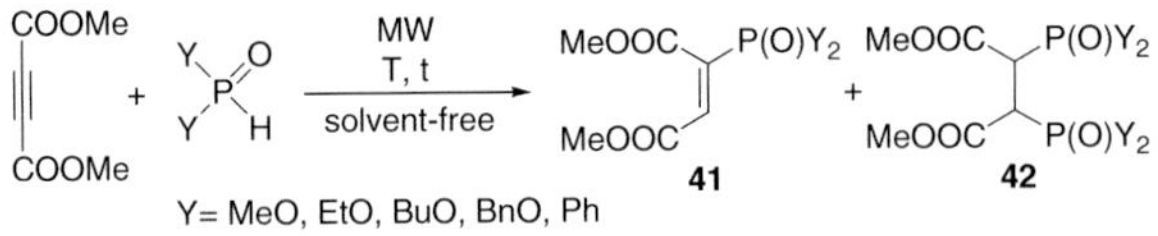

Y= MeO, EtO, BuO, BnO, Ph

**Scheme 3.29** MW-assisted addition of dialkyl phosphites and diphenylphosphine oxide to dimethyl acetylenedicarboxylate

**Table 3.7** The MW-assisted addition of dialkyl phosphites and diphenylphosphine oxide to dimethyl acetylenedicarboxylate

| Entry | Y | $\frac{n\,Y_2P(O)H}{n(MeO_2CC)_2}$ | T (°C) | t (h) | Product composition (%) | | Yield (%) |
|---|---|---|---|---|---|---|---|
| | | | | | **41** | **42** | |
| 1 | MeO | 0.5 | 90 → 100 | 5.5 | 55 | 45 | 45 (**41** (Y = MeO)) |
| 2 | MeO | 2 | 100 | 3.5 | 5 | 95 | 90 (**42** (Y = MeO)) |
| 3 | EtO | 0.5 | 90 → 100 | 5.5 | 51 | 49 | 46 (**41** (Y = EtO)) |
| 4 | EtO | 2 | 100 | 3.5 | 7 | 93 | 87 (**42** (Y = EtO)) |
| 5 | BuO | 0.5 | 90 → 100 | 5.5 | 50 | 50 | 44 (**41** (Y = BuO)) |
| 6 | BuO | 2 | 100 | 3.5 | 6 | 94 | 90 (**42** (Y = BuO)) |
| 7 | BnO | 0.5 | 90 → 100 | 5.5 | 43 | 57 | 39 (**41** (Y = BnO)) |
| 8 | BnO | 2 | 100 | 3.5 | 0 | 100 | 96 (**42** (Y = BnO)) |
| 9 | Ph | 0.5 | 26 | 0.25 | 58 | 42 | 53 (**41** (Y = Ph)) |
| 10 | Ph | 2 | 80 | 0.75 | 4 | 96 | 94 (**42** (Y = Ph)) |

The similar reaction of alkyl phenylpropiolates and two equivalents of dialkyl phosphites at 190 °C afforded a mixture of *E* and *Z* alkyl 3-(dialkoxyphosphoryl)-3-phenylacrylates (**43**), and in a few cases some of the bisadducts (**44**). Within the phosphoryl-phenylacrylates (**43**) the *E* isomer predominated (Scheme 3.30, Table 3.8) [62].

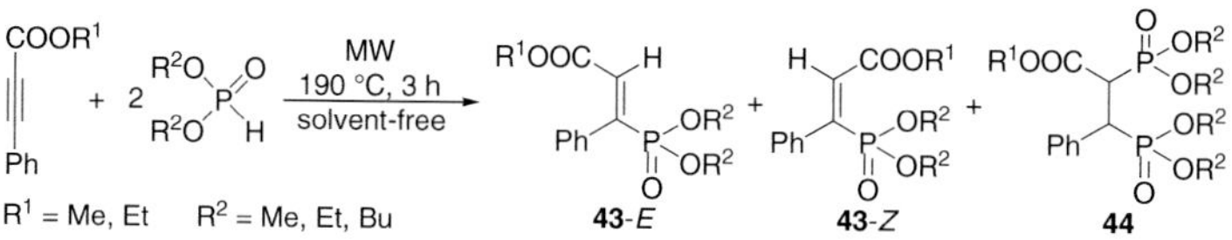

**Scheme 3.30** MW-assisted addition of dialkyl phosphites to methyl- and ethyl phenylpropiolate

**Table 3.8** The MW-assisted addition of dialkyl phosphites to alkyl phenylpropiolates

| Entry | $R^1$ | $R^2$ | Product composition (%) | | | Yield (43-E) (%) |
|---|---|---|---|---|---|---|
| | | | Monoadduct (43) | | Bisadduct (44) | |
| | | | E | Z | | |
| 1 | Me | Me | 73 | 13 | 14 | 60 |
| 2 | Me | Et | 90 | 10 | 0 | 84 |
| 3 | Me | Bu | 89 | 11 | 0 | 83 |
| 4 | Et | Me | 74 | 14 | 12 | 66 |
| 5 | Et | Et | 86 | 14 | 0 | 80 |
| 6 | Et | Bu | 85 | 15 | 0 | 72 |

## 3.8 The Addition of >P(O)H Species to Carbonyl-Compounds

α-Aryl-α-hydroxyphosponates and α-aryl-α-hydroxyphosphine oxides (**45**), potentially bioactive substrates, were synthesized in a catalytic and solvent-free MW-assisted reaction comprising the addition of >P(O)H species to aryl aldehydes (Scheme 3.31, Table 3.9) [63].

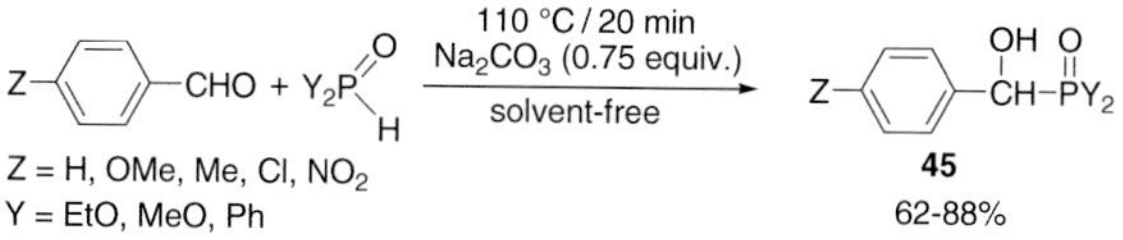

**Scheme 3.31** MW-assisted synthesis of α-hydroxyphosphonates and α-hydroxyphosphine oxides

**Table 3.9** Summary of the MW-assisted synthesis of α-hydroxyphosphonates and α-hydroxyphosphine oxides

| Entry | Y | Z | Yield of **45** (%) |
|-------|-----|-----|---------------------|
| 1 | EtO | H | 85 |
| 2 | MeO | H | 87 |
| 3 | Ph | H | 88 |
| 4 | EtO | MeO | 82 |
| 5 | MeO | MeO | 84 |
| 6 | Ph | MeO | 78 |
| 7 | EtO | Me | 87 |
| 8 | MeO | Me | 62 |
| 9 | Ph | Me | 80[a] |
| 10 | EtO | Cl | 84 |
| 11 | MeO | Cl | 72 |
| 12 | Ph | Cl | 79[a] |
| 13 | EtO | NO$_2$ | 86[b] |
| 14 | MeO | NO$_2$ | 71 |
| 15 | Ph | NO$_2$ | 80 |

[a]110 °C, 0.5 h
[b]150 °C, 1 h

Dialkyl phosphites were also added to the carbonyl function of α-ketophosphonates (**46**) to result in the formation of dronate analogue α-hydroxybisphosphonates (**47**) in the presence of diethylamine and in the absence of any solvent. Under optimum conditions, the formation of the rearranged by-product **48** could be avoided (Scheme 3.32) [64, 65].

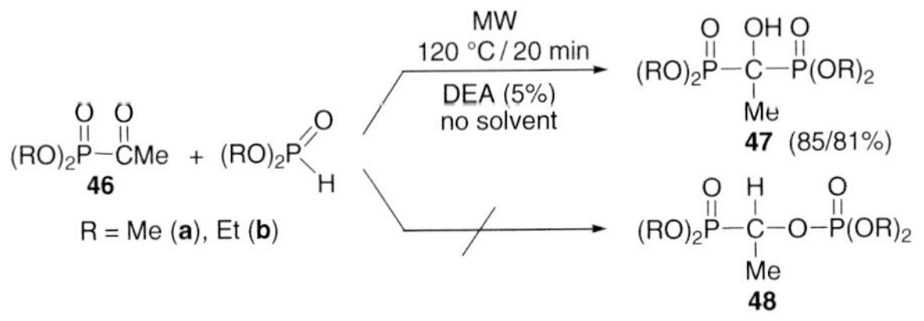

**Scheme 3.32** MW-assisted reaction of α-ketophosphonates and dialkyl phosphites

It was interesting to find that on MW irradiation, α-ketophosphonate **46b** was converted to α-hydroxybisphosphonate **47b**. Half of the starting material (**46b**) served as the precursor for diethyl phosphite which then reacted with the unchanged **46b** to afford bisphosphonate **27b** (Scheme 3.33) [65].

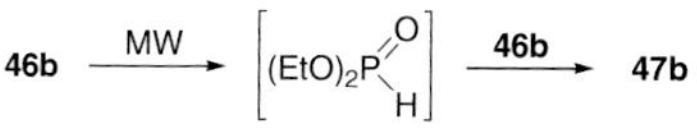

$$46b \xrightarrow{\text{MW}} \left[ (EtO)_2P \overset{O}{\underset{H}{\diagdown}} \right] \xrightarrow{46b} 47b$$

**Scheme 3.33** Preparation of hydroxybisphosphonate **47b** directly from the α-ketophosphonate precursor

## 3.9 The Conversion of α-Hydroxyphosphonates to α-Aminophosphonates

Surprisingly, α-hydroxyphosphonates **25** could be easily transformed to the corresponding α-aminophosphonates (**49**) by reaction with primary amines (Scheme 3.34) [66]. The substitution reaction was promoted by the neighbouring group effect of the adjacent P=O function.

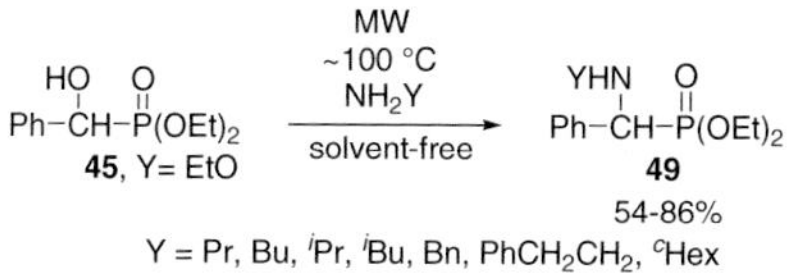

$$\underset{\substack{\textbf{45, Y= EtO}}}{\overset{\substack{HO \quad O \\ || \\ Ph-CH-P(OEt)_2}}{}} \xrightarrow[\substack{\text{solvent-free}}]{\substack{MW \\ \sim100\ °C \\ NH_2Y}} \underset{\textbf{49}}{\overset{\substack{YHN \quad O \\ || \\ Ph-CH-P(OEt)_2}}{}}$$

54-86%

Y = Pr, Bu, $^i$Pr, $^t$Bu, Bn, PhCH$_2$CH$_2$, $^c$Hex

**Scheme 3.34** Preparation of α-aminophosphonates by substitution reaction of α-hydroxyphosphonates

## 3.10 The Kabachnik–Fields (Phospha-Mannich) Reaction

α-Aminophosphonates (**50**, Y = RO), the analogues of α-aminoacid esters, and α-aminophosphine oxides (**50**, Y = Ph) were synthesized by the solvent- and catalyst-free MW-assisted Kabachnik–Fields condensation of primary amines, aldehydes/ketones and >P(O)H reagents, such as dialkyl phosphites and diphenylphosphine oxide. Earlier preparations utilized special catalysts (e.g. BiNO$_3$ [67], phthalocyanine [68], and Lantanoid(OTf)$_3$ [69]), which mean cost and environmental burden. It was found that under MW conditions there is no need for any catalyst. Moreover, the condensation was performed in the absence of any solvent (Scheme 3.35, Table 3.10) [70]. In a few cases, the >P(O)H species was added to the preformed Schiff base (R$^1$N=CR$^2$R$^3$).

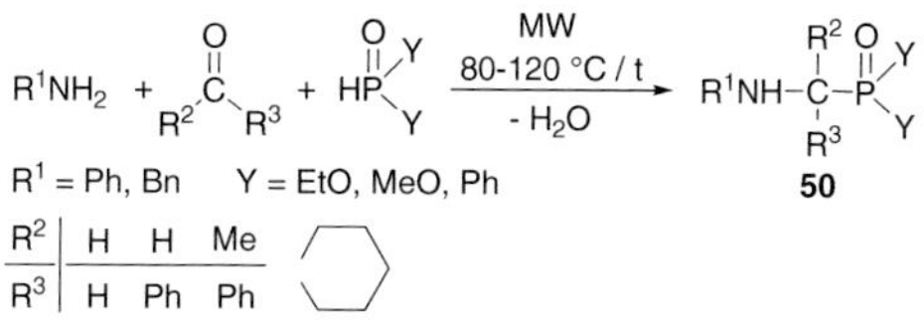

$$R^1NH_2 + \underset{\substack{R^2 \quad R^3}}{\overset{O}{\underset{||}{C}}} + HP\overset{O}{\underset{Y}{\diagdown}}_Y \xrightarrow[\substack{-H_2O}]{\substack{MW \\ 80-120\ °C\,/\,t}} \underset{\textbf{50}}{R^1NH-\underset{\substack{R^3}}{\overset{\substack{R^2}}{\underset{||}{C}}}-P\overset{O}{\underset{Y}{\diagdown}}_Y}$$

R$^1$ = Ph, Bn   Y = EtO, MeO, Ph

| R$^2$ | H | H | Me | ⬡ |
|-------|---|---|----|---|
| R$^3$ | H | Ph | Ph | |

**Scheme 3.35** MW-assisted Kabachnik–Fields condensations

**Table 3.10** Summary of the MW-assisted Kabachnik–Fields reactions

| Entry | $R^1$ | $R^2$ | $R^3$ | Y | T (°C) | t (min) | Yield of **50** (%) |
|-------|-------|-------|-------|-----|--------|---------|---------------------|
| 1 | Ph | H | H | EtO | 80[a] 100[b] | 15[a] 30[b] | 91 |
| 2 | Ph | H | H | MeO | 80[a] 80[b] | 15[a] 60[b] | 80 |
| 3 | Ph | H | H | Ph | 80 | 30 | 94 |
| 4 | Bn | H | H | EtO | 100 | 30 | 81 |
| 5 | Bn | H | H | Ph | 80 | 30 | 88 |
| 6 | Ph | H | Ph | EtO | 100 | 30 | 93 |
| 7 | Ph | H | Ph | MeO | 100 | 30 | 86 |
| 8 | Ph | H | Ph | Ph | 80 | 30 | 87 |
| 9 | Bn | H | Ph | EtO | 100 | 20 | 83 |
| 10 | Bn | H | Ph | MeO | 100 | 20 | 87 |
| 11 | Ph | Me | Ph | EtO | 120 | 40 | 80 |
| 12 | Bn | Me | Ph | EtO | 120 | 30 | 84 |
| 13 | Bn | Me | Ph | Ph | 100[a] 120[b] | 30[a] 30[b] | 80 |
| 14 | Ph | ⬡ | | EtO | 120 | 40 | 81 |
| 15 | Bn | ⬡ | | EtO | 120 | 30 | 91 |
| 16 | Bn | ⬡ | | MeO | 120 | 30 | 85 |
| 17 | Bn | ⬡ | | Ph | 100[a] 120[b] | 30[a] 30[b] | 80 |

[a]Condensation of the oxo-component and the amine
[b]Addition of the >P(O)H species to the Schiff-base

MW-assisted phospha-Mannich condensations were also performed in an excess of diethyl phosphite. Due to the use of domestic MW ovens, the reaction temperatures were not reported [71, 72].

The use of heterocyclic amines pyrrolidine, piperidine derivatives, morpholine and piperazine derivatives or heterocyclic >P(O)H species (e.g. 1,3,2-dioxaphosphorine oxide) led to N-heterocyclic [73] and P-heterocyclic [74] α-aminophosphonates (**51**, Y = EtO and **52**) and α-aminophosphine oxides (**51**, Y = Ph) (Schemes 3.36 and 3.37).

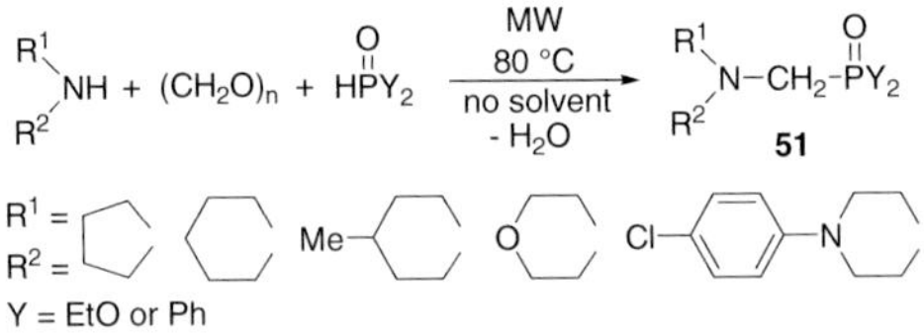

**Scheme 3.36** Kabachnik–Fields reactions with N-heterocycles as the amine component

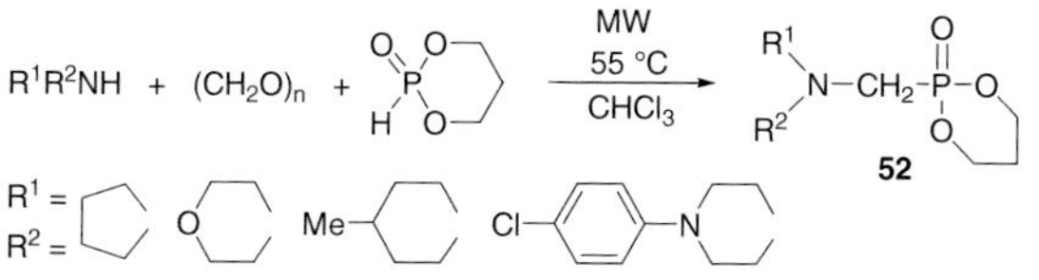

**Scheme 3.37** Kabachnik–Fields reactions with 1,3,2-dioxaphosphorine oxide as the P-reactant

In the reaction of dialkylamines, paraformaldehyde and dibenzo[c.e][1,2]-oxaphosphorine oxide (**53**), the heterocyclic ring underwent ring opening by reaction with water formed in the condensation to result in end-product **55** (Scheme 3.38) [74].

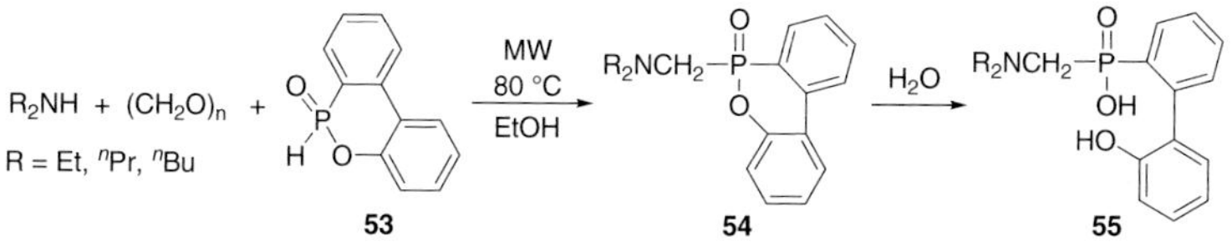

**Scheme 3.38** Kabachnik–Fields reaction with a dibenzooxaphosphorine oxide as the P-reactant

Then 3-amino-6-methyl-2H-pyran-2-ones (**56**) were utilized in the Kabachnik–Fields reaction with formaldehyde and dialkyl phosphites or diphenylphosphine oxide (Scheme 3.39) [75].

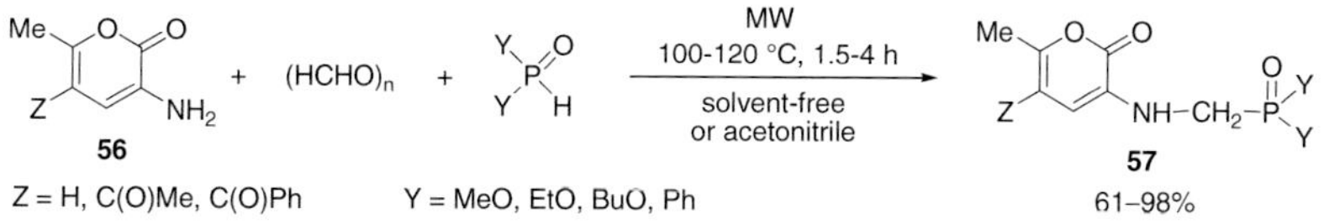

**Scheme 3.39** The synthesis of phosphono- or phosphinoylmethylamino-2H-pyran-2-ones

Phospha-Mannich-condensations are also known to proceed with trialkyl phosphites in water as the solvent. In this respect, the reaction of benzylamine, benzaldehyde and triethyl phosphite was investigated in comparison with the variation using diethyl phosphite as the P-reagent. The first version was practically complete at room temperature, but the conversion with diethyl phosphite remained uncomplete (Scheme 3.40) [76]. This experience justifies again the MW-assisted and solvent-free accomplishment of the Kabachnik–Fields reactions utilizing dialkyl phosphites [70].

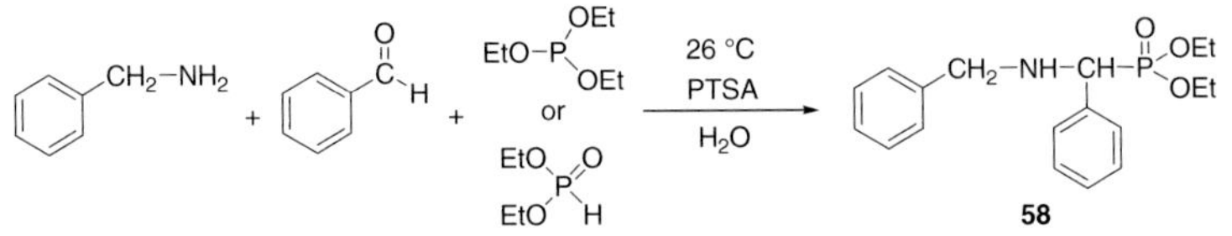

**Scheme 3.40** Kabachnik–Fields condensation with triethyl phosphite or diethyl phosphite in water

Primary amines are able to participate in bis(Kabachnik–Fields) condensations [77]. In such cases, alkyl or arylamines were reacted with two equivalents of the formaldehyde and the >P(O)H species to afford the bis($Z^1Z^2$P(O)CH$_2$)amines (**59**) (Scheme 3.41) [78–80]. Most of the reactions could be carried out without the use of any solvent, but for example the conversions with diphenylphosphine oxide had to be performed in acetonitrile due to the heterogeneity.

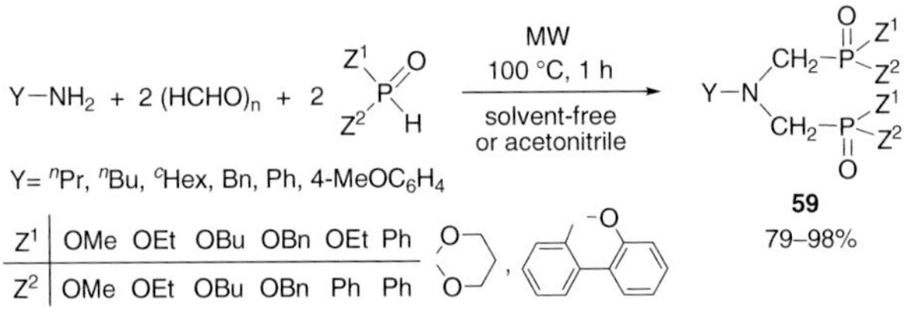

**Scheme 3.41** The bis(Kabachnik–Fields) reaction

The bisphosphinoyl derivatives (**59**, $Z^1=Z^2=$Ph) were transformed after double-deoxygenation to bis(phosphines) that were useful in the synthesis of ring platinum complexes [79, 80].

α-, β- and γ-amino acids (or esters) (**60**) were also utilized in the double Kabachnik–Fields condensation to furnish the bis(phosphono- or phosphinoyl) products (**61**) (Scheme 3.42) [81, 82].

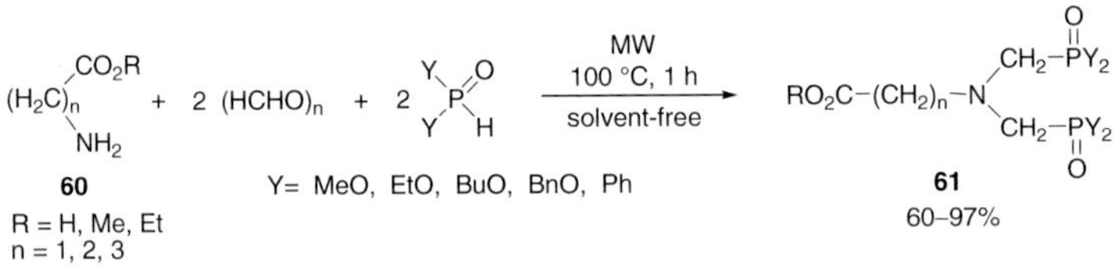

**Scheme 3.42** Bis(Kabachnik–Fields) reactions with amino acid derivatives

As further bis(Kabachnik–Fields) reactions, paraphenylene diamine was reacted with two equivalents of benzaldehyde derivatives and triethyl phosphite or diethyl phosphite (Scheme 3.43/(1)), or terephthalaldehyde was reacted with two equivalents of arylamine and P-reagent (as above) (Scheme 3.43/(2)). The bis-products **62** and **63** were obtained in variable yields [83].

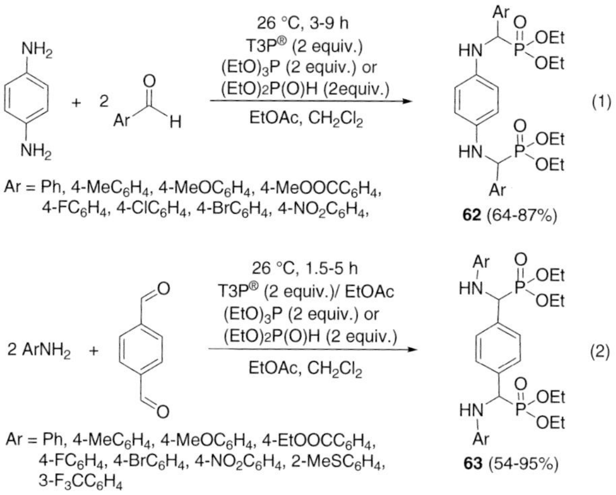

26 °C, 3-9 h
T3P® (2 equiv.)
(EtO)3P (2 equiv.) or
(EtO)2P(O)H (2equiv.)
EtOAc, CH2Cl2
Ar = Ph, 4-MeC6H4, 4-MeOC6H4, 4-MeOOCC6H4,
4-FC6H4, 4-ClC6H4, 4-BrC6H4, 4-NO2C6H4,
62 (64-87%)
(1)
26 °C, 1.5-5 h
T3P® (2 equiv.)/ EtOAc
(EtO)3P (2 equiv.) or
(EtO)2P(O)H (2 equiv.)
EtOAc, CH2Cl2
2 ArNH2 +
Ar = Ph, 4-MeC6H4, 4-MeOC6H4, 4-EtOOCC6H4,
4-FC6H4, 4-BrC6H4, 4-NO2C6H4, 2-MeSC6H4,
3-F3CC6H4
63 (54-95%)
(2)


**Scheme 3.43** Further variations of the bis(phospha-Mannich) reactions

## 3.11 Inverse Wittig-Type Reactions

The use of the MW technique was rather advantageous in the inverse Wittig-type reaction of 2,4,6-triisopropylphenyl-3-phospholene oxides, 2,4,6-triisopropylphenyl-phospholane oxides and 2,4,6-triisopropylphenyl-1,2-dihydrophosphinine oxides (all represented by formula **64**) and dimethyl acetylenedicarboxylate to furnish β-oxophosphoranes **65** (Scheme 3.44). Completion under thermal conditions required a *ca.* 2 week's heating at 150 °C, while on MW irradiation, the reactions were complete already after 3 h at the same temperature. No solvent had to be used in either case [84, 85].

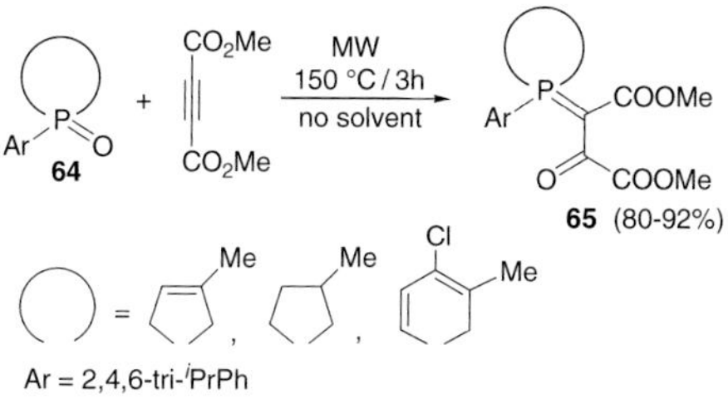

64
MW
150 °C / 3h
no solvent
65 (80-92%)
Ar = 2,4,6-tri-iPrPh


**Scheme 3.44** The inverse Wittig-type reaction of P-aryl substituted cyclic phosphine oxides with dimethyl acetylenedicarboxylate

It should be noted that in case of the *P*-mesityl substituent (Ar = 2,4,6-Me₃Ph), the inverse Wittig-type reaction shown in Scheme 3.44 took place exclusively under MW irradiation.

## 3.12   Diels–Alder Cycloaddition Reactions

The reaction of 1-phenyl-1,2-dihydrophosphinine oxide **66** with dienophiles, such as *N*-phenylmaleimide and dimethyl acetylenedicarboxylate took place according to the [4+2] Diels–Alder protocol to provide the corresponding phosphabicyclo[2.2.2]octene oxide (**67**) or phosphabicyclo[2.2.2]octadiene oxide (**68**), respectively [86]. The MW technique was useful in shortening the reaction times [30 min (MW) versus 2 days (thermal heating) under solvent-free conditions] and in providing the cycloadducts (**67/68**) in almost quantitative yields (Scheme 3.45) [87].

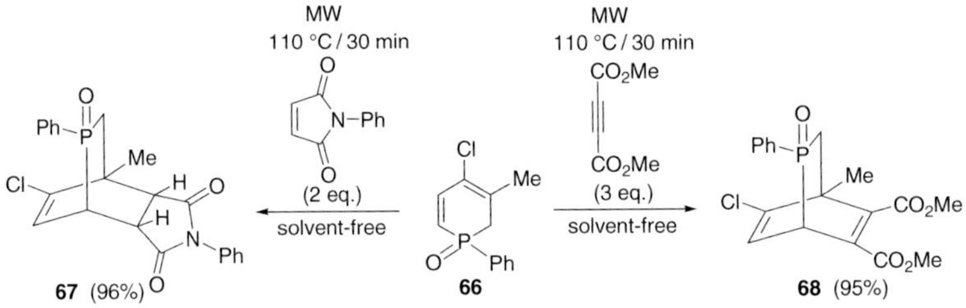

**Scheme 3.45**  Diels–Alder reactions of a 1,2-dihydrophosphinine oxide with different dienophiles

The absorption of MW irradiation was more efficient in the presence of onium salts, when a solvent was also used [88].

## 3.13   Fragmentation-Related Phosphorylations

The bridged *P*-heterocycles, such as phosphabicyclo[2.2.2]octadiene oxide **68** are useful in fragmentation-related phosphorylations [86]. On thermal influence or photochemical irradiation, the bridging methylenephosphine oxide [PhP(O)(CH₂)] unit of precursor **68** is ejected, and may phosphorylate a nucleophile e.g. a phenol added to the mixture prior to the fragmentation [89]. An alternative mechanism comprising a pentacoordinate intermediate was also proposed and proved [86]. It was found that the fragmentation-related phosphorylations are more efficient under MW conditions using ionic liquids as the solvent. The phosphorylated phenols (**69**) were obtained in somewhat better yields than in earlier experiments (Scheme 3.46) [90].

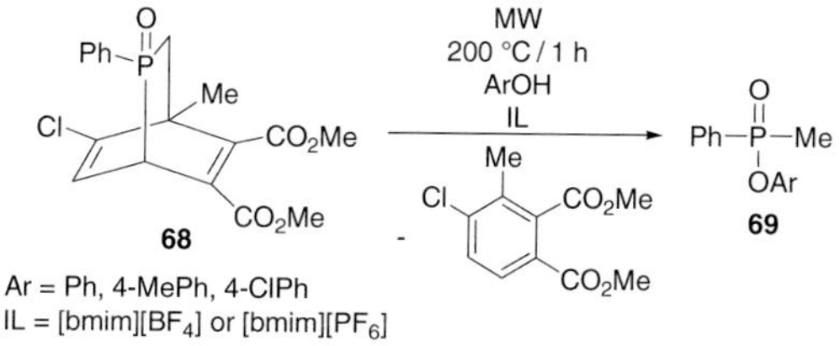

**Scheme 3.46** Fragmentation-related phosphorylations utilizing phosphabicyclo[2.2.2]octadiene precursor

## 3.14   Conclusions

In summary, a number of organophosphorus reactions were performed under MW conditions. The transformations studied embraced esterifications, amidations, alcoholyses, cycloadditions, additions, substitutions and condensations. In most cases, the role of MW irradiation was to make the reactions faster, or to make them more efficient and selective. In a few instances, the reactions took place only under MW conditions. It also occurred that MW irradiation substituted for the catalyst, or simplified catalyst systems. The examples shown demonstrate the potential of MW irradiation in organophosphorus synthesis.

**Acknowledgements** Supports from the Hungarian Research Development and Innovation Fund (K119202) and the Hungarian Scientific Research Fund (PD111895) are gratefully acknowledged.

## References

1. Guenin E, Meziane D (2011) Microwave assisted phosphorus organic chemistry: A review. Curr Org Chem 15:3465–3485. doi:10.2174/138527211797374724
2. Keglevich G, Grün A, Bálint E, Kiss NZ, Jablonkai E (2013) Microwave-assisted organophosphorus synthesis. Curr Org Chem 17:545–554. doi:10.2174/1385272811317050009
3. Keglevich G, Greiner I (2014) The meeting of two disciplines: organophosphorus and green chemistry. Curr Green Chem 1:2–16. doi:10.2174/2213346101011312181094831
4. Keglevich G, Kiss NZ, Mucsi Z, Jablonkai E, Bálint E (2014) The synthesis of phosphinates: traditional versus green chemical approaches. Green Process Synth 3:103–110. doi:10.1515/gps-2013-0106
5. Keglevich G (2014) Microwave-assisted synthesis of P-heterocycles. Phosphorus, Sulfur Silicon Relat Elem 189:1266–1278. doi:10.1080/10426507.2014.885974
6. Keglevich G, (2015) Application of microwave irradiation in the synthesis of P-heterocycles. In: Brahmachari G (ed) Green synthetic approaches for biologically relevant heterocycles. Elsevier, Amsterdam, pp 559–570. doi:10.1016/B978-0-12-800070-0.00020-7

7. Keglevich G, Grün A, Bagi P, Bálint E, Kiss N, Kovács R, Jablonkai E, Kovács T, Fogassy E, Greiner I (2015) Environmentally friendly chemistry with organophosphorus syntheses in focus. Per Polytechn Chem Eng 59:82–95. doi:10.3311/PPch.7317

8. Keglevich G, Novak T, Vida L, Greiner I (2006) Microwave irradiation as an alternative to phase transfer catalysis in the liquid-solid phase, solvent-free C-alkylation of active methylene containing substrates. Green Chem 8:1073–1075. doi:10.1039/B610481A

9. Keglevich G, Majrik K, Vida L, Greiner I (2008) Microwave irradiation as a green alternative to phase transfer catalysis: solid-liquid phase alkylation of active methylene containing substrates under solvent-free conditions. Lett Org Chem 5:224–228. doi:10.2174/157017808783955754

10. Greiner I, Grün A, Ludányi K, Keglevich G (2011) Solid–liquid two-phase alkylation of tetraethyl methylenebisphosphonate under microwave irradiation. Heteroatom Chem 22:11–14. doi:10.1002/hc.20648

11. Keglevich G, Grün A, Blastik Z, Greiner I (2011) Solid–liquid phase alkylation of P=O–functionalized CH acidic compounds utilizing phase transfer catalysis and microwave irradiation. Heteroatom Chem 22:174–179. doi:10.1002/hc.20673

12. Grün A, Blastik Z, Drahos L, Keglevich G (2012) Microwave-assisted alkylation of diethyl ethoxycarbonylmethylphosphonate under solventless conditions. Heteroat Chem 23:241–246. doi:10.1002/hc.21009

13. Keglevich G, Grün A (2015) Microwave irradiation as a substitute for phase transfer catalyst in C-alkylation reactions. Curr Green Chem 3:254–263. doi:10.2174/221334610266614121521218

14. Grün A, Bálint E, Keglevich G (2015) Solid–liquid phase C-alkylation of active methylene containing compounds under microwave conditions. Catalysts 5:634–652. doi:10.3390/catal5020634

15. Keglevich G, Grün A, Bálint E (2013) Microwave irradiation and phase transfer catalysis in C-, O- and N-alkylation reactions. Curr Org Synth 10:751–763. doi:10.2174/1570179411310050006

16. Grün A, Blastik Z, Drahos L, Keglevich G (2014) Dialkylation of diethyl ethoxycarbonylmethylphosphonate under microwave and solventless conditions. Heteroat Chem 25:107–113. doi:10.1002/hc.21142

17. McBride JJ Jr, Grange L, Mais A (1963) Preparing esters of phosphinic acids. USA Patent US3092650, 04/20/1961

18. Quin LD (2000) A guide to organophosphorus chemistry. Wiley, New York

19. Kiss NZ, Keglevich G (2014) An overview of the synthesis of phosphinates and phosphinic amides. Curr Org Chem 18:2673–2690. doi:10.2174/1385272819666140829011741

20. Keglevich G, Kiss NZ, Mucsi Z (2014) Synthesis of phosphinic acid derivatives; traditional versus up-to-date synthetic procedures. Chem Sci J 5:1–14. doi:10.4172/2150-3494.1000088

21. Kiss NZ, Ludányi K, Drahos L, Keglevich G (2009) Novel synthesis of phosphinates by the microwave-assisted esterification of phosphinic acids. Synth Commun 39:2392–2404. doi:10.1080/00397910802654880

22. Troev KD (2006) Reactivity of H-phosphonates. In: Chemistry and application of H-phosphonates. Elsevier, Amsterdam, pp 23–105. doi:10.1016/B978-044452737-0/50004-1

23. Keglevich G, Kiss NZ, Körtvélyesi T, Mucsi Z (2013) Direct esterification and amidation of phosphinic acids under microwave conditions. Phosphorus, Sulfur Silicon Relat Elem 188:29–32. doi:10.1080/10426507.2012.743542

24. Keglevich G, Bálint E, Kiss NZ, Jablonkai E, Hegedűs L, Grün A, Greiner I (2011) Microwave-assisted esterification of phosphinic acids. Curr Org Chem 15:1802–1810. doi:10.2174/138527211795656570

25. Keglevich G, Kiss NZ, Mucsi Z, Körtvélyesi T (2012) Insights into a surprising reaction: The microwave-assisted direct esterification of phosphinic acids. Org Biomol Chem 10:2011–2018. doi:10.1039/C2OB06972E

26. Kiss NZ, Böttger É, Drahos L, Keglevich G (2013) Microwave-assisted direct esterification of cyclic phosphinic acids. Heteroat Chem 24:283–288. doi:10.1002/hc.21092

27. Mucsi Z, Kiss NZ, Keglevich G (2014) A quantum chemical study on the mechanism and energetics of the direct esterification, thioesterification and amidation of 1-hydroxy-3-methyl-3-pholene 1-oxide. RSC Adv 4:11948–11954. doi:10.1039/C3RA47456A

28. Keglevich G, Kiss NZ, Drahos L, Körtvélyesi T (2013) Direct esterification of phosphinic acids under microwave conditions: extension to the synthesis of thiophosphinates and new mechanistic insights. Tetrahedron Lett 54:466–469. doi:10.1016/j.tetlet.2012.11.054

29. Jablonkai E, Henyecz R, Milen M, Kóti J, Keglevich G (2014) T3P®-assisted esterification and amidation of phosphinic acids. Tetrahedron 70:8280–8285. doi:10.1016/j.tet.2014.09.021

30. Jablonkai E, Milen M, Drahos L, Keglevich G (2013) Esterification of five-membered cyclic phosphinic acids under mild conditions using propylphosphonic anhydride (T3P®). Tetrahedron Lett 54:5873–5875. doi:10.1016/j.tetlet.2013.08.082

31. Bálint E, Jablonkai E, Bálint M, Keglevich G (2010) Alkylating esterification of 1-hydroxy-3-pholene oxides under solventless MW conditions. Heteroat Chem 21:211–214. doi:10.1002/hc.20596

32. Keglevich G, Bálint E, Karsai É, Grün A, Bálint M, Greiner I (2008) Chemoselectivity in the microwave-assisted solvent-free solid–liquid phase benzylation of phenols: O- versus C-alkylation. Tetrahedron Lett 49:5039–5042. doi:10.1016/j.tetlet.2008.06.051

33. Keglevich G, Bálint E, Karsai É, Varga J, Grün A, Bálint M, Greiner I (2009) Heterogeneous phase alkylation of phenols making use of phase transfer catalysis and microwave irradiation. Lett Org Chem 6:535–539. doi:10.2174/157017809789869500

34. Jablonkai E, Bálint E, Balogh GT, Drahos L, Keglevich G (2012) Cyclic phosphinates by the alkylation of a thermally unstable 1-hydroxy-1,2-dihydrophosphinine 1-oxide and a 3-hydroxy-3-phosphabicyclo[3.1.0]hexane 3-oxide. Phosphorus, Sulfur Silicon Relat Elem 187:357–363. doi:10.1080/10426507.2011.613876

35. Bálint E, Tajti Á, Drahos L, Ilia G, Keglevich G (2013) Alcoholysis of dialkyl phosphites under microwave conditions. Curr Org Chem 17:555–562. doi:10.2174/1385272811317050010

36. Keglevich G, Bálint E, Tajti Á, Mátravölgyi B, Balogh György T, Bálint M, Ilia G (2014) Microwave-assisted alcoholysis of dialkyl phosphites by ethylene glycol and ethanolamine. Pure Appl Chem 86:1723–1728. doi:10.1515/pac-2014-0601

37. Keglevich G, Kiss NZ, Körtvélyesi T (2013) Microwave-assisted functionalization of phosphinic acids; amidations versus esterifications. Heteroat Chem 24:91–99. doi:10.1002/hc.21068

38. Kiss NZ, Simon A, Drahos L, Huben K, Jankowski S, Keglevich G (2013) Synthesis of 1-amino-2,5-dihydro-1H-phosphole 1-oxides and their N-phosphinoyl derivatives, bis (2,5-dihydro-1H-phoshol-1-yl)amine P, P'-dioxides. Synthesis 45:199–204. doi:10.1055/s-0032-1316830

39. Kiss NZ, Rádai Z, Mucsi Z, Keglevich G (2015) Synthesis of bis(phosphinoyl)amines and phosphinoyl-phosphorylamines by the N-phosphinoylation and N-phosphorylation of 1-alkylamino-2,5-dihydro-1H-phosphole 1-oxides. Heteroat Chem 26:134–141. doi:10.1002/hc.21229

40. Jablonkai E, Keglevich G (2014) P-C bond formation by coupling reaction utilizing >P(O)H species as the reagents. Curr Org Synth 11:429–453. doi:10.2174/15701794113109990066

41. Jablonkai E, Keglevich G (2014) Advances and new variations of the Hirao reaction. Org Prep Proc Int 46:281–316. doi:10.1080/00304948.2014.922376

42. Jablonkai E, Keglevich G (2013) P-Ligand-free, microwave-assisted variation of the Hirao reaction under solvent-free conditions; the P-C coupling reaction of >P(O)H species and bromoarenes. Tetrahedron Lett 54:4185–4188. doi:10.1016/j.tetlet.2013.05.111

43. Keglevich G, Jablonkai E, Balázs LB (2014) A "green" variation of the Hirao reaction: the P-C coupling of diethyl phosphite, alkyl phenyl-H-phosphinates and secondary phosphine oxides with bromoarenes using a P-ligand-free Pd(OAc)$_2$ catalyst under microwave and solvent-free conditions. RSC Adv 4:22808–22816. doi:10.1039/C4RA03292F

44. Amaya T, Abe Y, Inada Y, Hirao T (2014) Synthesis of self-doped conducting polyaniline bearing phosphonic acid. Tetrahedron Lett 55:3976–3978. doi:10.1016/j.tetlet.2014.04.115

45. Jablonkai E, Balázs L, Keglevich G (2015) A P-ligand-free nickel-catalyzed variation of the Hirao reaction under microwave conditions. Curr Org Chem 19:197–202. doi:10.2174/1385272819666150114235413

46. Kalek M, Ziadi A, Stawinski J (2008) Microwave-assisted palladium-catalyzed cross-coupling of aryl and vinyl halides with H-phosphonate diesters. Org Lett 10:4637–4640. doi:10.1021/ol801935r

47. Andaloussi M, Lindh J, Sävmarker J, Sjöberg PJR, Larhed M (2009) Microwave-promoted palladium(II)-catalyzed C-P bond formation by using arylboronic acids or aryltrifluoroborates. Chem Eur J 15:13069–13074. doi:10.1002/chem.200901473

48. Villemin D, Jaffrès P-A, Siméon F (1997) Rapid and efficient phosphonation of aryl halides catalysed by palladium under microwaves irradiation. Phosphorus, Sulfur Silicon Relat Elem 130:59–63. doi:10.1080/10426509708033697

49. Rummelt SM, Ranocchiari M, van Bokhoven JA (2012) Synthesis of water-soluble phosphine oxides by Pd/C-catalyzed P–C coupling in water. Org Lett 14:2188–2190. doi:10.1021/ol300582y

50. Jablonkai E, Keglevich G (2015) Catalyst-free P–C coupling reactions of halobenzoic acids and secondary phosphine oxides under microwave irradiation in water. Tetrahedron Lett 56:1638–1640. doi:10.1016/j.tetlet.2015.02.015

51. Jablonkai E, Keglevich G (2015) A survey of the palladium–catalyzed Hirao reaction with emphasis on green chemical aspects. Curr Green Chem 2:379–391. doi:10.2174/2213346102999150630114117

52. Jablonkai E, Keglevich G (2015) P–C coupling reactions under environmentally-friendly conditions. In: Petrova V (ed) Advances in engineering research, vol 10. Nova Science Publishers Inc, pp 99–125

53. Dzielak A, Mucha A (2015) Catalytic and MW-assisted Michaelis-Arbuzov reactions. Curr Green Chem 2:223–236. doi:10.2174/2213346102666150128195001

54. Keglevich G, Grün A, Bölcskei A, Drahos L, Kraszni M, Balogh GT (2012) Synthesis and proton dissociation properties of arylphosphonates; a microwave-assisted catalytic Arbuzov reaction with aryl bromides. Heteroat Chem 23:574–582. doi:10.1002/hc.21053

55. Keglevich G, Sipos M, Takács D, Greiner I (2007) A study on the Michael addition of dialkylphosphites to methylvinylketone. Heteroat Chem 18:226–229. doi:10.1002/hc.20266

56. Keglevich G, Sipos M, Takács D, Ludányi K (2008) Phospha-Michael reactions involving P-heterocyclic nucleophiles. Heteroat Chem 19:288–292. doi:10.1002/hc.20421

57. Keglevich G, Sipos M, Imre T, Ludányi K, Szieberth D, Tőke L (2002) Diastereoselective synthesis of 1,2,3,6-tetrahydrophosphinine 1-oxides with an exocyclic P-function by a Michael type addition. Tetrahedron Lett 43:8515–8518. doi:10.1016/S0040-4039(02)02081-6

58. Keglevich G, Sipos M, Szieberth D, Nyulászi L, Tm Imre, Ludányi K, Tőke L (2004) Weak intramolecular interactions as controlling factors in the diastereoselective formation of 3-phosphinoxido- and 3-phosphono-1,2,3,6-tetrahydrophosphinine 1-oxides. Tetrahedron 60:6619–6627. doi:10.1016/j.tet.2004.05.090

59. Jablonkai E, Drahos L, Drzazga Z, Pietrusiewicz KM, Keglevich G (2012) 3-P(O)< Functionalized phospholane 1-oxides by the Michael reaction of 1-phenyl-2-phospholene 1-oxide and dialkyl phosphites, H-phosphinates or diphenylphosphine oxide. Heteroat Chem 23:539–544. doi:10.1002/hc.21047

60. Bálint E, Takács J, Drahos L, Keglevich G (2012) Microwave-assisted phospha-Michael addition of dialkyl phosphites, a phenyl-H-phosphinate, and diphenylphosphine oxide to maleic derivatives. Heteroat Chem 23:235–240. doi:10.1002/hc.21007

61. Keglevich G, Bálint E, Takács J, Drahos L, Huben K, Jankowski S (2014) The addition of dialkyl phosphites and diphenylphosphine oxide on the triple bond of dimethyl acetylenedicarboxylate under solvent-free and microwave conditions. Curr Org Synth 11:161–166. doi:10.2174/1570179411999140304142747

62. Bálint E, Takács J, Bálint M, Keglevich G (2015) The catalyst-free addition of dialkyl phosphites on the triple bond of alkyl phenylpropiolates under microwave conditions. Curr Catal 4:57–64. doi:10.2174/2211544704666150303232225
63. Keglevich G, Róza Tóth V, Drahos L (2011) Microwave-assisted synthesis of α-hydroxy-benzylphosphonates and -benzylphosphine oxides. Heteroat Chem 22:15–17. doi:10.1002/hc.20649
64. Grün A, Molnár IG, Bertók B, Greiner I, Keglevich G (2009) Synthesis of α-hydroxy-methylenebisphos-phonates by the microwave-assisted reaction of α-oxophosphonates and dialkyl phosphites under solventless conditions. Heteroat Chem 20:350–354. doi:10.1002/hc.20558
65. Keglevich G, Grün A, Molnár IG, Greiner I (2011) Phenyl-, benzyl-, and unsymmetrical hydroxy-methylenebisphosphonates as dronic acid ester analogues from α-oxophosphonates by microwave-assisted syntheses. Heteroat Chem 22:640–648. doi:10.1002/hc.20727
66. Kiss NZ, Kaszás A, Drahos L, Mucsi Z, Keglevich G (2012) A neighbouring group effect leading to enhanced nucleophilic substitution of amines at the hindered α-carbon atom of an α-hydroxyphosphonate. Tetrahedron Lett 53:207–209. doi:10.1016/j.tetlet.2011.11.026
67. Bhattacharya A, Kaur T (2007) An efficient one-pot synthesis of alpha-amino phosphonates catalyzed by bismuth nitrate pentahydrate. Synlett 5:745–748. doi:10.1002/chin.200730153
68. Matveeva ED, Podrugina TA, Tishkovskaya EV, Tomilova LG, Zefirov NS (2003) A novel catalytic three-component synthesis (Kabachnick-Fields reaction) of alpha-aminophosphonates from ketones. Synlett 2003:2321–2324. doi:10.1055/s-2003-42118
69. Lee S, Park J, Kang J, Lee J (2001) Lanthanide triflate-catalyzed three component synthesis of alpha-amino phosphonates in ionic liquids. A catalyst reactivity and reusability study. Chem Commun:1698–1699. doi:10.1016/j.tetlet.2005.12.027
70. Keglevich G, Szekrényi A (2008) Eco-friendly accomplishment of the extended Kabachnik–Fields reaction; a solvent- and catalyst-free microwave-assisted synthesis of α-aminophosphonates and α-aminophosphine oxides. Lett Org Chem 5:616–622. doi:10.2174/157017808786857598
71. Kabachnik MM, Zobnina EV, Beletskaya IP (2005) Catalyst-free microwave-assisted synthesis of alpha-aminophosphonates in a three-component system: (RC)-C-1(O)R-2-(EtO)(2)P(O)H-RNH₂. Synlett 2005:1393–1396. doi:10.1055/s-2005-868519
72. Mu X-J, Lei M-Y, Zou J-P, Zhang W (2006) Microwave-assisted solvent-free and catalyst-free Kabachnik-Fields reactions for α-amino phosphonates. Tetrahedron Lett 47:1125–1127. doi:10.1016/j.tetlet.2005.12.027
73. Prauda I, Greiner I, Ludányi K, Keglevich G (2007) Efficient synthesis of phosphono- and phosphinoxidomethylatcd N-heterocycles under solvent-free microwave conditions. Synth Commun 37:317–322. doi:10.1080/00397910601033856
74. Keglevich G, Szekrényi A, Sipos M, Ludányi K, Greiner I (2008) Synthesis of cyclic aminomethylphosphonates and aminomethyl-arylphosphinic acids by an efficient microwave-mediated phospha-Mannich approach. Heteroat Chem 19:207–210. doi:10.1002/hc.20387
75. Bálint E, Takács J, Drahos L, Juranović A, Kočevar M, Keglevich G (2013) α-Aminophosphonates and α-aminophosphine oxides by the microwave-assisted Kabachnik-Fields reactions of 3-amino-6-methyl-2H-pyran-2-ones. Heteroat Chem 24:221–225. doi:10.1002/hc.21086
76. Keglevich G, Bálint E, Kangyal R, Bálint M, Milen M (2014) A critical overview of the Kabachnik-Fields reactions utilizing trialkyl phosphites in water as the reaction medium; a study on the benzaldehyde-benzylamine triethyl phosphite/diethyl phosphite models. Heteroat Chem 25:282–289. doi:10.1002/hc.21192
77. Cherkasov RA, Garifzyanov AR, Talan AS, Davletshin RR, Kurnosova NV (2009) Synthesis of new liophilic functionalized aminomethylphosphine oxides and their acid-base and membrane-transport properties toward acidic substrates. Russ J Gen Chem 79:1835–1849. doi:10.1134/S1070363209090114

78. Keglevich G, Szekrényi A, Szöllősy Á, Drahos L (2011) Synthesis of bis (phosphonatomethyl)-, bis(phosphinatomethyl)-, and bis(phosphinoxidomethyl)amines, as well as related ring bis(phosphine) platinum complexes. Synth Commun 41:2265–2272. doi:10.1080/00397911.2010.501478

79. Bálint E, Fazekas E, Pintér G, Szőllősy A, Holczbauer T, Czugler M, Drahos L, Körtvélyesi T, Keglevich G (2012) Synthesis and utilization of the bis(>P(O)CH2)amine derivatives obtained by the double Kabachnik–Fields reaction with cyclohexylamine; quantum chemical and X-ray study of the related bidentate chelate platinum complexes. Curr Org Chem 16:547–554. doi:10.2174/138527212799499822

80. Bálint E, Fazekas E, Pongrácz P, Kollár L, Drahos L, Holczbauer T, Czugler M, Keglevich G (2012) N-Benzyl and N-aryl bis(phospha-Mannich adducts): synthesis and catalytic activity of the related bidentate chelate platinum complexes in hydroformylation. J Organomet Chem 717:75–82. doi:10.1016/j.jorganchem.2012.07.031

81. Bálint E, Fazekas E, Drahos L, Keglevich G (2013) The synthesis of N, N-bis (dialkoxyphosphinoylmethyl)- and N, N-Bis(diphenylphosphinoylmethyl)glycine esters by the microwave-assisted double Kabachnik–Fields reaction. Heteroat Chem 24:510–515. doi:10.1002/hc.21126

82. Bálint E, Fazekas E, Kóti J, Keglevich G (2015) Synthesis of N, N-bis (dialkoxyphosphinoylmethyl)- and N, N-bis(diphenylphosphinoylmethyl)-β- and γ-amino acid derivatives by the microwave-assisted double Kabachnik–Fields reaction. Heteroat Chem 26:106–115. doi:10.1002/hc.21221

83. Milen M, Ábrányi-Balogh P, Kangyal R, Dancsó A, Frigyes D, Keglevich G (2014) T3P®-mediated one-pot synthesis of bis(α-aminophosphonates). Heteroat Chem 25:245–255. doi:10. 1002/hc.21170

84. Keglevich G, Dudás E, Sipos M, Lengyel D, Ludányi K (2006) Efficient synthesis of cyclic β-oxophosphoranes by the microwave-assisted reaction of cyclic phosphine oxides and dialkyl acetylenedicarboxylate. Synthesis 2006:1365–1369. doi:10.1055/s-2006-926395

85. Keglevich G, Forintos H, Körtvélyesi T (2004) Synthesis and reactions of β-oxophosphoranes/ylides containing a cyclic or acyclic P-moiety. Curr Org Chem 8:1245–1261. doi:10.2174/1385272043370023

86. Keglevich G, Szelke H, Kovács J (2004) Fragmentation-related phosphinylation and phosphonylation of nucleophiles utilising the bridging P-unit of 2-phosphabicyclo[2.2.2] oct-5-ene derivatives. Curr Org Synth 1:377–389. doi:10.2174/1570179043366521

87. Keglevich G, Dudás E (2007) Microwave promoted efficient synthesis of 2-phosphabicyclo [2.2.2]octadiene- and octene 2-oxides under solvent-free conditions in Diels–Alder reaction. Synth Commun 37:3191–3199. doi:10.1080/00397910701547532

88. Hohmann E, Keglevich G, Greiner I (2007) The effect of onium salt additives on the Diels–Alder reactions of a 1-phenyl-1,2-dihydrophosphinine oxide under microwave conditions. Phosphorus, Sulfur Silicon Relat Elem 182:2351. doi:10.1080/10426500701441473

89. Keglevich G, Szelke H, Dobó A, Nagy Z, Toke L (2001) Phosphorylation of phenols and naphthols by phenyl-methylenephosphine oxide generated by the thermolysis of a 2-phosphabicyclo[2.2.2]octa-57-diene 2-oxide. Synth Commun 31:1737–1741. doi:10.1081/ scc-100104403

90. Keglevich G, Kovács R, Drahos L (2011) Diels–Alder cycloadditions of 1,2-dihydrophosphinine oxides and fragmentation-related phosphorylations with 2-phosphabicyclo[2.2.2]octadiene oxides under green chemical conditions—the role of microwave and ionic liquids. Phosphorus, Sulfur Silicon Relat Elem 186:2172–2179. doi:10.1080/10426507.2011.597807

# Chapter 4
# Interpretation of the Effects of Microwaves

Péter Bana and István Greiner

**Abstract** During the past three decades of microwave (MW) assisted organic chemistry, the initial observations of unexpected reaction behavior obtained in MW reactors grew to a general understanding of MW effects. This chapter aims to present the currently accepted theories of MW rate enhancements, and a few of the main steps leading to today's understanding of these phenomena. Modern experimental techniques in MW chemistry revealed the fundamental role of temperature in interpreting the outcome of MW heated experiments. However, temperature can be realized on different spatial scales, which will be used as the basis of our classification. This way, the phenomena associated with MW heating are differentiated between macroscopic and microscopic effects, both of which will be discussed in detail.

**Keywords** Microwave chemistry · Dielectric heating · Macroscopic thermal effects · Thermal effects in heterogeneous systems · Temperature measurement · Microscopic scale thermal microwave effects · Microscopic thermal alterations · Microwave-actuated reactions · Non-thermal microwave effects

## 4.1 Introduction

The concept of microwave (MW) activation of organic reactions, the so-called "MW effects" is nearly the same age as the chemical use of MW heating and the field of MW chemistry itself. When a MW heated reaction led to results (altered reaction rate, yield and product distribution) that were unparalleled by the conventionally heated counterpart, speculative rationalizations began on the underlying

P. Bana (✉)
Department of Organic Chemistry and Technology, Budapest University of Technology and Economics, 1521 Budapest, Hungary
e-mail: bana.peter@mail.bme.hu

I. Greiner
Gedeon Richter Plc, PO Box 27, 1475 Budapest, Hungary

© Springer International Publishing Switzerland 2016
G. Keglevich (ed.), *Milestones in Microwave Chemistry*,
SpringerBriefs in Green Chemistry for Sustainability,
DOI 10.1007/978-3-319-30632-2_4

physical and chemical processes. As the field of MW chemistry and the involved experimental techniques gradually matured, the concepts of MW effects were reiterated several times. This chapter aims to present the currently accepted theories of MW rate enhancements, and a few of the main steps leading to today's understanding of these phenomena.

## 4.1.1  Early Observations and Rationalizations

In their pioneering works published in 1986, the groups of Gedye [1] and Giguere [2] reported several examples of different reaction types (amide hydrolysis, esterification, oxidation, substitution, Diels-Alder reactions, Claisen rearrangements, and "ene" reaction) placed into a household (kitchen) MW oven that showed beneficial effects (improvements in reaction rates) compared to the conventionally heated control experiments. As the MW heated reactions were run in sealed vessels under pressure, the reaction temperatures were significantly higher than the boiling point of the solvent, which explains higher rates [3].

These studies not only marked the birth of MW chemistry, they also stimulated the search for further rate enhancements in MW heated experiments. As the number of examples rose in the early 1990s, such specific cases were found, where rate accelerations were realized at the same apparent temperature. In a MW assisted Diels–Alder reaction (Scheme 4.1), 8-fold decrease in reaction half-time ($t_{1/2}$) was observed leading to the cycloadduct (1), while temperature (measured by fiber optic thermometry) was similar to the conventionally heated control experiments [4]. The modification of the free energy of activation ($\Delta G^{\#}$) was proposed as the basis of rate enhancement, which was later re–interpreted [5] as the effect of localized "hot-spots" or superheating.

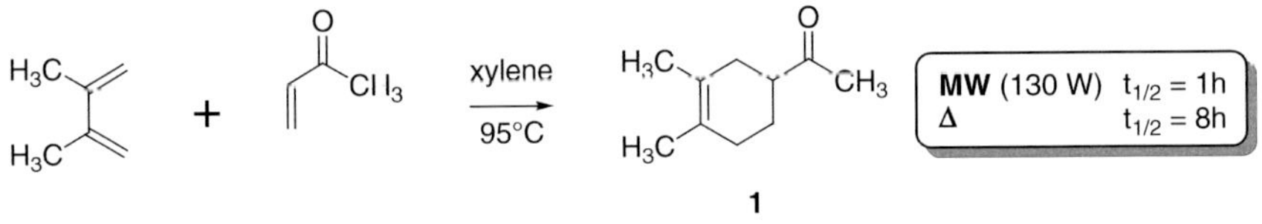

Scheme 4.1  Diels-Alder reaction of 2,3-dimethyl-butadiene and methyl-vinyl-ketone

Nevertheless, some other early claims of MW specific activation couldn't be reproduced by the independent re-investigations [6, 7], which also took experimental error into account [8]. These studies introduced the basic and (to the present day) most useful tool of MW chemistry: the comparison studies against conventionally heated control experiments, while apparent temperature of the reaction mixtures is kept at the same value. This methodology provides kinetically interpretable data, which can be used to draw the conclusions on the nature of reaction rate improvements.

## 4.1.2 Classification of the Phenomena

In order to organize the large body of diverse phenomena that was realized in the last three decades of MW chemistry, the effects of MW heating were classified in some distinct categories. The notion of MW-specific effects [7, 9, 10] was introduced to describe observations, when the outcome of the MW heated reaction was thought to be specifically connected to features of the dielectric heating process.

Now it is understood that temperature plays fundamental role when interpreting the outcome of MW heated experiments. Thus, the phenomena should be categorized, based on how the temperature is involved in the appropriate explanations. After some iteration, the most up to date classification was given by Kappe, which differentiates between three categories: *thermal effects* (based on differences of the bulk temperature), *specific MW effects* (temperature based changes that are uniquely connected to MW heating and cannot be demonstrated by conventional heating) and *non-thermal MW effects* (changes that cannot be rationalized by either thermal or specific MW effects) [11, 12].

In our opinion, this system addresses most of the experimentally observed phenomena, but it is inconsistent in some cases of specific effects. The distinction between specific MW effects and purely thermal effects is not always clear, because both use temperature as the main factor to determine reaction rate, but this can be interpreted on various temporal and spatial ranges. Consequently, some reviews [13, 14] categorize these observations as thermal effects. However, the difference in scale is fundamental from the practitioners' point of view, as the basis of the rationalization of the experimental outcome has to be the appropriately defined and measured temperature value.

We propose a revised system, in which the scale, on which temperature is realized is used as basis of classification: proportions from macroscopic (bulk), through local and microscopic (molecular) sizes will be considered. Only in cases, where temperature based considerations cannot be applied on any of these size ranges, is the concept of non-thermal effects used.

In this system, thermal effects are used to describe rate accelerations caused by elevated temperature of the reaction mixture itself, either in the bulk phase, locally or on a microscopic scale. In all of these cases, the interaction of MW field and matter results exclusively in heating. The alterations in reaction rate ($k$) can be rationalized [15] by the well-known Arrhenius kinetic equation (Eq. 4.1). Only the temperature ($T$) is affected, reaction mechanism is unaltered, and the reaction specific kinetic parameters [the pre-exponential factor ($A$) and the activation energy term ($E_a$)] are identical to the values measured under conventional heating.

$$k = Ae^{-\frac{E_a}{RT}} \qquad (4.1)$$

### 4.1.3 Theory of MW Dielectric Heating

In MW assisted organic synthesis, reaction mixtures are subjected to electromagnetic waves of 2450 MHz frequency, which results in absorption of electromagnetic energy by the sample, and its conversion into heat [10, 12, 15–18]. MW dielectric heating is a consequence of the action of the electric component of the electromagnetic field on the dielectric (non-conducting) material, through two fundamental mechanisms [15, 16].

*Dipolar polarization mechanism* describes interaction of the MW field with materials consisting of substances bearing dipole moment. The dipoles align themselves to the alternating electric field with a phase difference resulting in oscillating rotation of the dipoles. In this process energy is lost and converted to heat due to molecular collisions. The other major heating mechanism is the *ionic conduction mechanism*, during which charged particles are displaced in the alternating direction of the electric field. This causes oscillating motion and collision with other molecules and leads to energy being transferred from the electric field to the sample as heat.

The amount of energy converted to heat is related to the so-called loss factor (or dissipation factor, loss tangent, *tan δ*), which is characteristic of the material at a given frequency and temperature [16]. The higher the loss tangent (*tan δ*), the more efficient absorption and more rapid heating occurs. Solvents can be characterized as high- (*tan δ* > 0.5; e.g., ethylene glycol, ethanol, dimethyl sulfoxide), medium- (*tan δ* 0.1–0.5; e.g., acetic acid, *N,N*-dimethyl formamide, water), and low MW-absorbing (*tan δ* < 0.1; e.g., acetonitrile, dichloromethane, hexane) [19].

The overall absorption character of the mixture determines the heating speed and final temperature, while the sample's thermal properties (heat capacity, conductivity) are commonly less dominant [20]. Reaction mixtures consisting of polar or ionic reactants and polar solvents are very efficiently heated [3, 21]. In case of ionic liquids, high temperatures can be attained within a few seconds [22]. In contrast, low MW-absorbing solvents show slower rates of heating when the same MW power is applied. It should be kept in mind that in these cases the vessel's absorption can contribute to the heating [12, 23], since borosilicate glass has a considerable *tan δ* of 0.001 [24]. However, even these solvents are applicable, when polar reagents (which are commonly encountered in organic synthesis), ionic additives [16, 25–30] or phase transfer catalysts [31] are present. Addition of inert susceptors [32, 33] or strongly MW absorbing co-solvents [34] also leads to faster heating. Passive heating elements [23, 35] and reaction vessels [36–38] made of highly absorbing materials[1] serve the same purpose.

---

[1]Commonly used materials for this purpose are Weflon[TM] or Carboflon[®] (graphite loaded forms of Teflon[®]) and silicon carbide (SiC).

## 4.2  Macroscopic Thermal Effects in Homogeneous Systems

MW reactors are essentially very efficient tools for heating reactions mixtures. Thus, the majority of MW assisted organic reactions are conducted at high temperatures that cannot be practically achieved by common laboratory methods. Moreover, heating can be applied rapidly and uniformly through the entire volume of the sample. These features grant a prominent place for MW methods in modern preparative practice, even without taking any MW-specific effects in account.

Modern MW reactors have the ability to run experiments in temperature control mode [12], holding the sample at a more-or-less constant nominal temperature throughout the reaction time.[2] In contrast to irradiation with constant MW power (in which case the final temperature is dependent on several factors), temperature control technique gives chemists the ability to compare dielectrically heated experiments to the conventional procedures at the same temperature.

The efficiency of MW heating allows chemists to heat reactions to unusually high temperatures (>200 °C) [39], which is aided by the ease of applying sealed vessel conditions. In pressurized containers (autoclave-type conditions) solvents can be heated[3] above their boiling points [40]. Dramatic differences can arise, when reactions that are performed under standard oil bath conditions (i.e., heating under reflux) are compared with high-temperature MW-heated processes. In case of reactions requiring several hours at the reflux temperature of the particular solvent, reaction time can be reduced to minutes or seconds using overheated solvents in a sealed vessel [41], as illustrated by the synthesis (Scheme 4.2) of 2-methylbenzimidazole (**2**) [42].

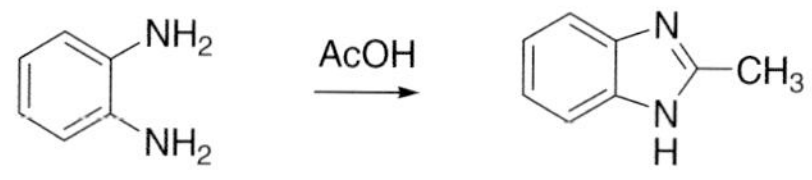

| Heating mode | Temperature | Reaction time |
|---|---|---|
| conventional | 60°C | 3 days |
| conventional | 100°C | 5 hours |
| MW | 130°C | 1 hour |
| MW | 200°C | 3 minutes |
| MW | 270°C | 1 second |

**Scheme 4.2**  Effect of temperature on the synthesis of 2-methylbenzimidazole

---

[2]This is achieved by the feedback control of the applied MW power, based on the signal of the applied temperature probe.

[3]The achievable temperature is limited by the volatility (vapor pressure at the target temperature) of the solvent and the pressure rating of the MW instrument. The pressure control mechanism is also critical to the maximal pressure value.

Caution should be taken, since preforming reactions at higher temperatures can give way to other reaction pathways, than the desired one. In a base-catalyzed multicomponent condensation [43], conventional heating at lower temperature provided the tricyclic Hantzsch-type dihydropyridine derivative (3), but the MW experiment heated to 150 °C introduced an alternative pathway, consisting of base-mediated ring opening—recyclization yielding an unanticipated tricyclic product (4) through a common intermediate (5) (Scheme 4.3).

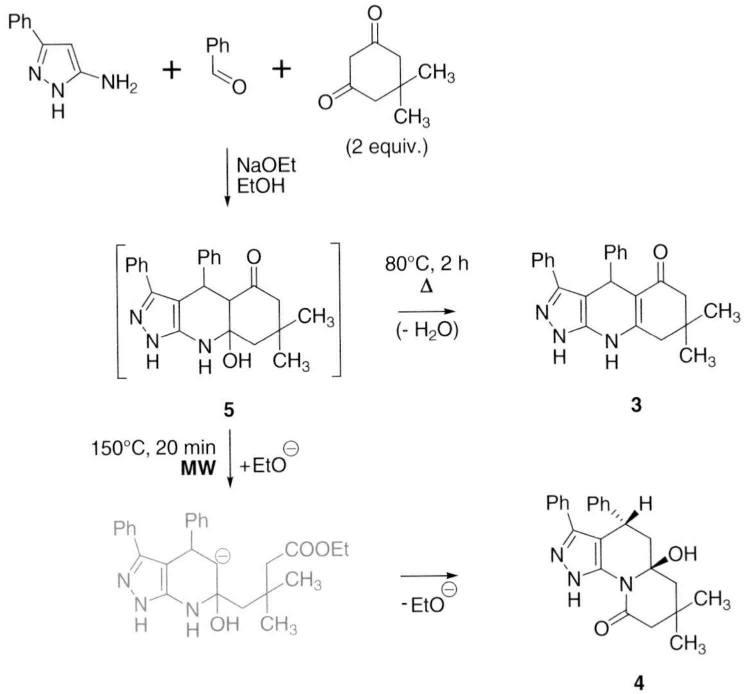

**Scheme 4.3** Three-component condensation of 5-aminopyrazoles, aromatic aldehydes, and diketones

The attainable high temperatures can lead to interesting behavior, when water is used as solvent [39, 44–47]. In fact, water has a dielectric constant that is strongly dependent on temperature, decreasing as the temperature is raised (from $\varepsilon_r = 78$ at 25 °C to $\varepsilon_r = 20$ at 300 °C), thus becoming a nearly MW transparent pseudo-organic solvent, which solvates organic molecules better (similarly to acetone). Furthermore, the dissociation constant of water increases by three orders of magnitude in the same temperature range, consequently it can act as acidic or basic catalyst.

Near-critical state water can also be used in MW assisted organic synthesis as a solvent in Fischer indole syntheses, pinacol rearrangements, Diels-Alder reactions, and as a reagent in hydrolysis and hydration reactions, where its strongly acidic properties provide acid catalysis as well [25].

In highly durable MW applicators (capable of withstanding 300 °C temperature and 80 bar pressure) supercritical conditions were reached in case of 1-butanol as solvent and utilized in a transesterification reaction of triglycerides with 1-butanol [48].

## 4.2.1  Effects of the Increased Rate of Heating

Owing to the increased rate of MW heating, the time required by heating-up can be significantly reduced. Moreover, after the set reaction time fast cooling can also be achieved in dedicated MW reactors. As a consequence, undesired processes taking place at temperatures below the temperature of the planned reaction (both before reaching the set temperature and during cooling) can be retarded, which leads to cleaner reaction mixtures and lower amounts of by-products [41, 49]. Product distribution can also be altered, if it is controlled by complex temperature-dependent kinetic profiles [50]. This often manifests in altered chemo- or regioselectivity [9, 51–53].

Even subtle thermal differences can considerably alter the outcome of a MW experiment, as it was emphasized in case of a palladium-catalyzed Buchwald-Hartwig reaction. Optimal activity of catalyst was limited in time, which could only be exploited by the rapid MW heating [54].

Change in the heating profile can influence the regioselectivity in the MW assisted sulfonation [55] of naphthalene (Scheme 4.4), which leads to mixtures containing lower amounts of 1-naphthalenesulfonic acid (6a), than expected in conventional heating. Different heating rates (achieved by changing the power setting) gave different selectivities, the ratio of 2-naphthalenesulfonic acid (6b) increased as the heating was more rapid [56]. The authors noted that isomeric ratio did not depend on the heating mode (i.e., MW or conventional), but only on the heating rate. It has been argued that rapid heating, leading to high temperatures favors the formation of the thermodynamically stable product, while conventional heating leads to the kinetically controlled product.

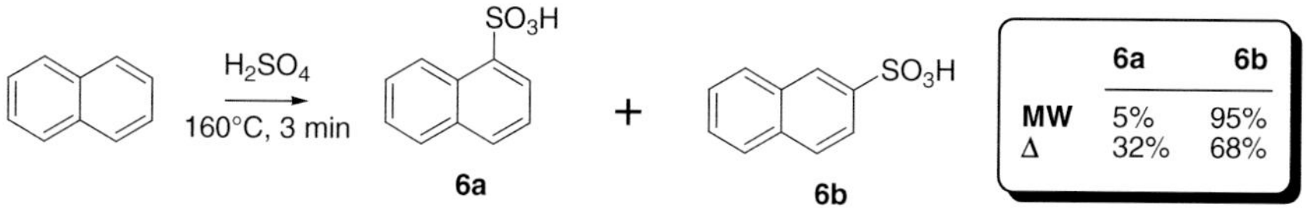

|        | 6a  | 6b  |
|--------|-----|-----|
| MW     | 5%  | 95% |
| Δ      | 32% | 68% |

**Scheme 4.4**  Sulfonation of naphthalene

Complete inversion in selectivity was observed in the benzylation of 2-pyridone. While conventional heating provides only the N-benzylated product MW heating leads to the mixture of C-benzylated products [57].

## 4.2.2  Effects of the Volumetric Heating

MW absorption processes occur uniformly in the dielectric material, resulting in simultaneous heating of the whole volume of the sample. In consequence, better temperature homogeneity can be achieved. Heat is directly transferred into the sample, in contrast to conventional heating that relies on heat transfer from the heated surface of the reaction flask by conduction and convection, which is typically slow, and necessarily results in a temperature gradient.

Volumetric heating is particularly useful, when high viscosity reaction mixtures are heated, or conduction and convection is retarded. It allows the application of solvent-free conditions, by eliminating the difficulties associated with heat and mass transfer [58–60]. In the transformation of urea to cyanuric acid (7) (Scheme 4.5), conventional heating is sluggish and leads to low yields, because the initially formed product (7) is a poor conductor of heat, and forms an insulating layer near the vessel walls. Increasing the temperature by conventional methods results in partial decomposition. In contrast, heating in a monomode MW device increases the temperature of the inner layers as well, and rapidly gives full conversion without side products [8].

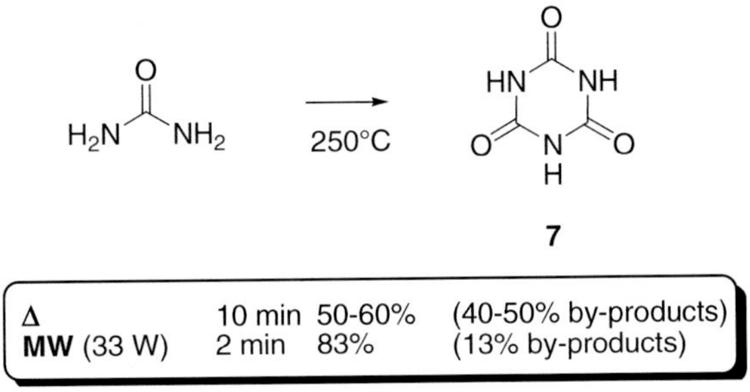

**Scheme 4.5**  Synthesis of cyanuric acid

| Δ | 10 min | 50-60% | (40-50% by-products) |
| MW (33 W) | 2 min | 83% | (13% by-products) |

The possibility to omit solvents, while homogenous heating is ensured, eliminates the restrictions of the solvent's boiling point on temperature, without the need for pressurized conditions. In the synthesis of 2-phosphabicyclo[2.2.2]octadiene-oxides (8) by Diels-Alder cycloaddition of 1,2-dihydrophosphinine oxides (9) and acetylenedicarboxylate (Scheme 4.6), the solvent-free conditions allowed by MW heating were beneficial, which was further enhanced by higher temperature [61].

Since heat is generated inside the sample, a thermal gradient orienting from the core to the vessel wall is exhibited, therefore the temperature profile (compared to

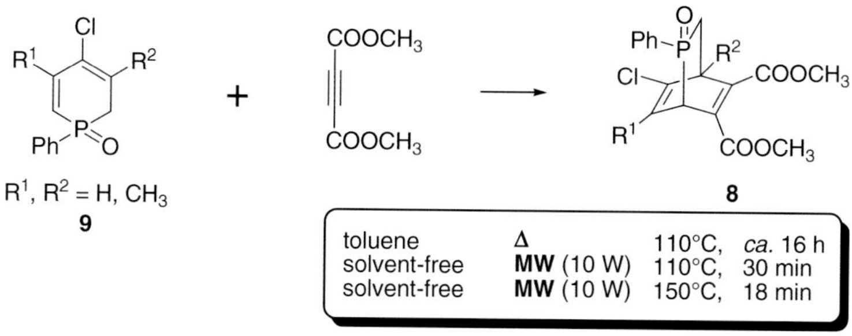

Scheme 4.6  Diels-Alder cycloaddition of 1,2-dihydrophosphinine oxides

conventional heating) is inverted [62]. Thermal imaging [63], temperature sensitive indicators and combined electro- and thermodynamic modeling [64] were used to compare temperature distributions to conductive heating. As the vessel wall is not heated, the so-called wall effects[4] are eliminated [39, 65]. However, a study on a ring closing metathesis reaction questioned the role of the wall effects in that case, as no difference was found, compared to the hot oil bath heated experiments [66].

### 4.2.3  Effects of Macroscopic Superheating at Atmospheric Pressure

A remarkable phenomenon can be observed, when liquids are kept in the state of boiling using MW heating under atmospheric conditions.[5] Solvents can be heated up to 38 °C above their normal boiling points [67–71], without applying sealed vessel or pressurized techniques. The elevated boiling temperature is explained by the inverted heat transfer, which is characteristic of MW heating [68], resulting in the lack of nucleation points on the vessel wall. Since nucleation is retarded and phase change requires higher energy [70], higher temperatures are needed to provide enough nucleation points for stable boiling.

This effect is dependent on the solvent, the overall geometry of the vessel and MW cavity and the MW power as well [69]. Different behavior can be observed in multimode and monomode MW systems [70]. Stirring [21], or adding boiling aids [71] as nucleation regulators prevents overheating. Furthermore, the elevated temperature is not constant in time. Consequently, it is difficult to reproduce this effect, so utilization is encumbered, although the possibility of this phenomenon should always be considered, when applying reflux conditions [72, 73].

---

[4]Wall effects are often harmful in conventional heating, e.g., leading to decomposition of products, catalyst deactivation.

[5]This is achieved by using a reflux condenser, similarly to the practice used in conventional heating.

   The magnitude of atmospheric superheating effect could be assessed using
photochemical reactions as molecular thermometers [74]. The Norrish type II
photochemical reaction of the mixture of valerophenones (**10a** and **10b**) was
conducted in different solvents (methanol, ethanol, t-butanol, acetonitrile) at reflux
conditions, using both conventional and dielectric heating. The photochemical
efficiency ratios (*R*), based on the ratio of fragmentation (**11a** and **11b**) *versus*
cyclization (**12a** and **12b**) was found to be linearly temperature dependent
(Scheme 4.7).

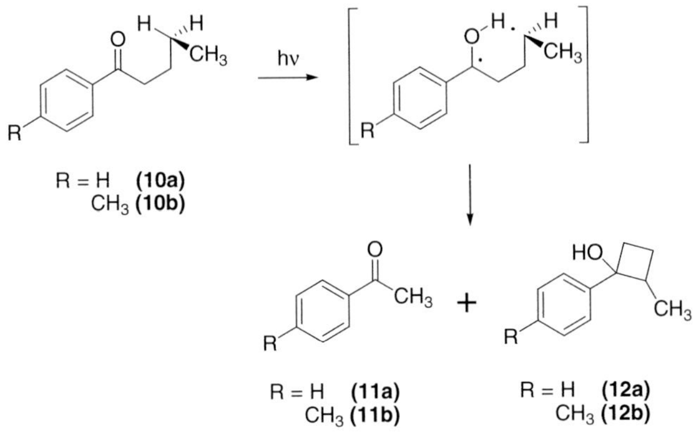

R = H    (**10a**)
CH$_3$ (**10b**)

R = H    (**11a**)
CH$_3$ (**11b**)

R = H    (**12a**)
CH$_3$ (**12b**)

**Scheme 4.7** Norrish type II photochemical reaction

   The temperature derived from this feature corresponded well with the bulk
temperature measured by internal fiber optic (FO) thermometry. The magnitude of
atmospheric superheating was estimated as 4–13 °C by both methods (Table 4.1).

**Table 4.1** Estimated superheating effects in the Norrish type II reaction

| Solvent | Slope of the correlation (1/°C) | Decrease of *R* (compared to the normal boiling point) | Superheating effect | |
|---|---|---|---|---|
| | | | Molecular thermometer (°C) | Fiber Optic thermometer (°C) |
| Methanol | 0.016 | 0.18 | 11 | 10 |
| Ethanol | 0.013 | 0.13 | 10 | 13 |
| *t*-butanol | 0.011 | 0.04 | 4 | 6 |
| Acetonitrile | 0.016 | 0.14 | 9 | 9 |

## 4.3   Thermal Effects in Heterogeneous Systems

The behavior of macroscopically inhomogeneous, multiphase systems in MW heating is fundamentally different from what is accustomed in heating by convection. Unlike the more or less uniform heat transfer characteristics of organic substances encountered in conventional heating, MW energy transfer depends on the much more diverse dielectric properties (determined by the molecular structure), which leads to significantly different behavior of chemically distinct phases of a heterogeneous system.

This way, differential heating is likely to occur in chemically inhomogeneous mixtures, as the loss tangent of the different phases are usually different. It was observed that the effects associated with these systems often cease to exist, when heterogeneity is eliminated by addition of co-solvent for instance [75, 76]. Heterogeneity itself can lead to changes in effective temperatures at the phase boundaries, as a result of interfacial MW polarization phenomena [77].

Real time in situ Raman spectroscopy was utilized to observe the phenomenon of "non-equilibrium local heating" occurring to the DMSO molecules in proximity of cobalt particles under MW irradiation. The observed local temperature anomaly contributed to the enhancement of reductive dehalogenation reactions on the catalyst surface [78]. Investigation of gold nanoparticles bearing thermally-labile fluorescent dyes under MW irradiation also provided evidence for localized overheating. At short distances (0.5 nm) from the surface, 70 °C excess temperature was evidenced, which rapidly diminished as distance from the particle increased [79].

Non-uniform temperature distribution was evidenced in zeolite-guest systems leading to altered selectivity in competitive sorption experiments [80]. Molecular dynamics simulations revealed non-equilibrium conditions caused by MW irradiation [81, 82].

The most practical significance is associated with the selective heating of well absorbing heterogeneous catalysts in low polarity reaction media [83]. Localized "hot-spots" were evidenced on the surface of the heterogeneous catalyst in gas phase reactions [84–86], resulting from its efficient heating. This causes high reaction rates on the catalyst, while the reactants and products situated in the cooler bulk phase are less likely exposed to thermal degradation. The macroscopic "hot-spots" in question were estimated to possess temperatures 100–200 °C higher than the bulk temperature, and size in the range of 90 μm up to 1000 μm [87]. In one study, the alterations of the catalyst surface after MW irradiation led to higher activity, which seemed to remain in subsequent conventionally heated experiments [88]. It should be noted that the observations are not general, in some cases the temperature of the catalyst didn't show significant increase [89].

In case of the MW accelerated hydrogenation reaction (Scheme 4.8) of a diene (13) to the saturated product (14) on heterogeneous palladium-on-charcoal (Pd/C) catalyst, the palladium metal is a strong MW absorber, while the solvent (ethyl acetate) has low *tan δ* value [90].

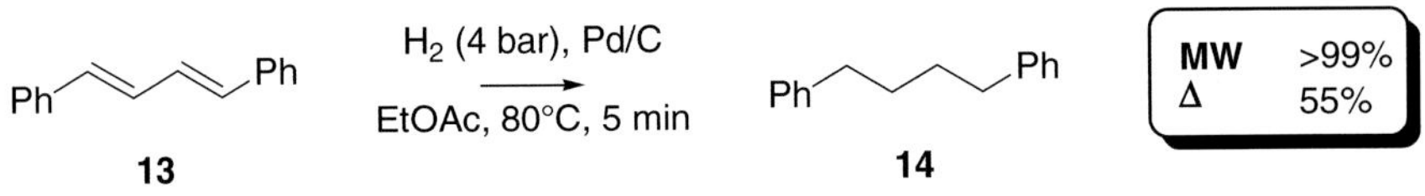

**Scheme 4.8** Catalytic hydrogenation of a diene

Other MW assisted catalytic reductions [91, 92] has also been developed. Selective heating was also suggested in palladium-catalyzed racemizations [93] and cross-coupling reactions [94, 95]. However, it should be noted that direct evidence for selective catalyst heating couldn't be obtained in this type of reactions, due to the complexity of the experimentally investigated systems [96, 97].

Heterogeneous chromium dioxide ($CrO_2$) based Magtrieve[TM] reagent was employed in MW assisted oxidation reactions [98, 99]. MW irradiation (200 W, 2 min) of a suspension of the reagent in MW transparent toluene resulted is "hot-spots" reaching *ca.* 140 °C temperature, revealed by thermal imaging. Exposing pure samples of the reagent to the same conditions led to surface temperatures of 360 °C.

Similar localized superheating or macroscopic "hot-spots" can also be expected in solid samples [100] heated in MW cavities. This leads to significant rate enhancements in the MW reaction of solid substrates often encountered in inorganic chemistry [101, 102], as well as organic synthesis conducted on solid support [103, 104]. Differential heating can also be exploited in such multiphase liquid/liquid systems, in which the sensitive product is transferred to the relatively cooler apolar organic phase, where decomposition of the sensitive product is less likely [105, 106].

Colloidal size inhomogeneities can also be considered as separate, differently heated phases [107]. Selective heating of highly absorbing precursors in MW transparent solvents led to exceptional results in the synthesis of CdSe and CdTe nanomaterials [108, 109]. Observations explainable by similar arguments were made for enzymatic reactions as well. In some cases optimal catalytic activity [110] or denaturation [111] were reported at far lower bulk temperatures than expected in case of conventional heating.

The importance of direct interaction between the electric field and magnesium metal was emphasized in the formation of Grignard reagents (**15**) under MW irradiation (Scheme 4.9). Significantly higher reaction rates were attributed to the visually observable arcing of the magnesium turnings resulting in freshly formed, clear surfaces on the magnesium metal (so-called electrostatic etching), where higher reactivity can be expected. Similar results could be obtained by adding an initiator (1,2-dibromoethane) to the reaction mixture [75, 112].

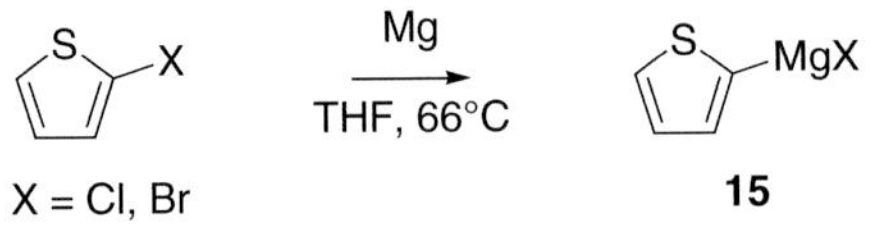

**Scheme 4.9** Grignard reagent formation from 2-halotiophenes without initiator

More interestingly, an exhaustive investigation of a closely related system showed that, while low field densities resulted in activation, applying higher density electromagnetic field at the same bulk temperature, the formation of the organometallic species was retarded. Arcing occurred with increased intensity in this case, causing the THF solvent to decompose, and cover the surface of the metal with a passivating layer, thus preventing the access of the reactant [113].

## 4.4   Consequences of MW Heating Beyond the Measurable Temperature

Generally speaking, all forms of the described thermal effects can be attributed to temperature, defined in a macroscopic scale that can be—in theory—demonstrated when using proper temperature measurement techniques [12, 22, 63, 114–118].

Apart from the aforementioned thermal effects, separate classes of phenomena are described, in which the macroscopically measurable temperature differences play no role in explaining the reaction rate alterations. This can either mean that rate difference stems from thermal variations that are microscopic in size, consequentially the commonly applied temperature measurement methods are unable to reveal them. On the other hand, one can imagine such interactions of MW and matter that are fundamentally different from the heating process caused by MW irradiation and whose effects on chemical reactions are independent from the bulk or local temperature.

Similarly to other empirical disciplines, more than one possible explanations can be devised for a particular observation. Both *microscopic thermal effects* and *non-thermal effects* of the MW irradiation manifest in similar experimental results (i.e., deviation from the thermally expected kinetics) and are usually indistinguishable from each other (based on the obtainable amount of data). In contrast, their theoretical origin is axiomatically different.

## 4.5   Thermal MW Effects on the Microscopic Scale

The concept of microscopic thermal effects is a special case of explanations that are based on thermal arguments. It is assumed that MW irradiation alters the temperature or the connected kinetic energy of smaller domains, leading to higher than the bulk temperature thereof that cannot be measured macroscopically. These effects are essentially different from bulk thermal phenomena, and can only be interpreted on a microscopic scale. This unique scale motivates us, to discuss these explanations separately from bulk thermal effects.

Microscopic thermal alterations can be most intuitively described as "hot-spots" or local overheatings, small regions of the reaction mixture, where the characteristic temperature (or average kinetic energy of the molecules) is higher than in the bulk phase. The concept of "hot-spots" was first applied to MW chemistry [4] as an analogy to ultrasound chemistry, where it was well established as the explanation [119] of the sonochemical effect of mechanical waves that generate specific sites of activation.

In some reviews, microscopic thermal effects are related to selective heating, which occur in macroscopically heterogeneous multiphase systems, and thus are categorized as specific effects [11, 12]. This way, the term "hot-spot" can not only refer to a differently heated phase in a heterogeneous system, but also to a microscopic region with altered temperature in a homogeneous phase. To avoid confusion it should be specified, whether a macroscopic or a microscopic "hot-spot" is discussed [83]. In this section microscopic "hot-spot" phenomena, occurring in homogeneous systems are discussed.

Microscopic thermal effects generally influence chemical reactions in the same basic way as bulk thermal effects. The higher temperature of the "hot-spots" results in accelerated reaction rate inside these regions, described by the Arrhenius equation. Besides the altered temperate term, the pre-exponential factor ($A$) and activation energy ($E_a$) are unaffected, these values are characteristic of the reaction and identical to the ones measured unconventionally heated experiments. Consequently, microscopic "hot-spots" provide a convenient way of assessing the rate accelerations in MW chemistry, by utilizing readily available kinetic data, in a way that is easily interpretable by chemists.

Since the local overheatings cannot be studied directly, their properties can only be investigated indirectly by measuring their effects on reaction rates. Most of our knowledge on their characteristic size, temperature and other features is based on estimation. For the same reason, their origin and the resulting kinetics within can be explained in different ways. However most of these theories are basically parallel to each other.

### 4.5.1 Characteristics of Microscopic Thermal Alterations

It is obvious that the "hot-spots" at hand represent a small fraction of the reaction mixture, while the remaining part is at the bulk (measurable) temperature. Theories suggest even distribution within the sample, and their size is generally thought to be much smaller than the characteristic dimensions of the reaction vessel, otherwise they would cause macroscopically identifiable inhomogeneities. Overheatings in "nano-size" range [120], domains having dimensions of 1–3 nm [121] and single solvent cages [122, 123] were described.

Temperature enhancements up to 60 °C associated with the overheated segments were suggested throughout the literature [19, 120, 124]. Direct temperature measurement on the microscopic scale was attempted by using Raman spectroscopy as a tool for thermometry on the molecular level (by exploiting the temperature dependence of intensities of particular molecular vibrations).Various polar molecules (benzaldehyde, chlorobenzene and $Cr(CO)_3(\eta^6\text{-}C_6H_5OMe)$ complex) dissolved in apolar solvents (hexane or diisopropyl ether) were investigated under varied MW power irradiation. However, no evidence was found for local overheating or other thermal alterations from temperature of the bulk reaction medium [125].

Based on the observed reaction rate, estimations of the elevated temperature inside the localized overheatings can be made. These studies also provide indirect evidence for such microscopic "hot-spots", by rationalizing the experimental results. The kinetic constants can be based either on measured or theoretical values. The volumetric concentration and relative temperature of the hypothesized "hot-spots" should also be taken into account, although in the absence of experimental data these parameters can only be chosen arbitrarily.

In an imidization reaction in polymer chemistry the temperature at the reaction site was approximated to be 50 °C higher than the observed temperature [126]. An early modeling study [100] used 2 % population of "hot-spots" with temperatures 70 °C above the bulk temperature. It was shown that such small density of superheating areas sufficiently leads to large rate accelerations, while its effect on the average temperature is minimal and virtually undetectable.

A recent quantitative model [120, 127] used an elaborate distribution of local temperature to explain the observed rate enhancement in the MW assisted reactions (Scheme 4.10) of a phosphonic acid (**16**). In these transformations, statistically occurring local overheating was found to be responsible for overcoming high enthalpy of activation [128, 129], thus the potential of MW heating in reactions with pronounced activation barriers was emphasized. More interestingly, while the slightly endothermic esterification lead to the product (**17a**) in high conversion, strong endothermic character prevented complete conversions in the syntheses of the thioester (**17b**) and the amide (**17c**) [130].

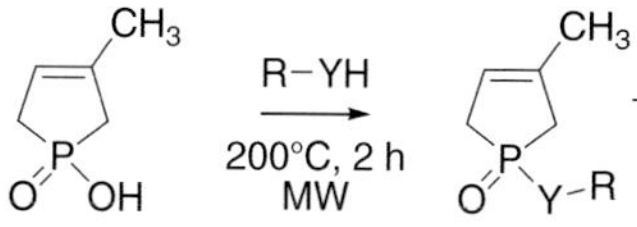

| Product | R–YH | $\Delta H^0$ (kJ/mol)* | $\Delta H^{\#}$ (kJ/mol)* | Conv. |
|---|---|---|---|---|
| 17a | nBu–OH | 0.8 | 134.6 | 58% |
| 17b | nBu–SH | 47.9 | 230.5 | 18% |
| 17c | Hex–NH$_2$ | 35.2 | 114.1 | 26% |

*calculated values (level of theory: B3LYP/6-31G(d,p))

**Scheme 4.10** Derivatization of a phosphonic acid

In an attempt to rationalize the results, several temperature models were considered, in which the overheated segment represented 5–30 % of the reaction volume. Within the overheated segments, temperature is exponentially distributed between discrete values of 5–50 °C. Using the kinetic data for the model reaction obtained from quantum chemical calculations, relative reaction rates could be calculated for the considered models of overheating. This could be compared to the experimental data of the synthesis of phosphonic ester (**17a**), to identify the best fitting model, in which overheating in 20 % of the reaction volume accounted for the observed rate enhancement under MW assisted conditions.

## 4.5.2  Origins of Microscopic Thermal Alterations

The physical causes of the localized overheating are rather unclear, the first integrated (physics, chemistry and experimental) studies were awaited till recent years.

Generation of "hot-spots" by molecular scale dielectric relaxation was suggested in the early literature [70]. The rate of MW energy transfer was also considered as a possible origin. It was argued that the characteristic time-frame associated with MW energy transfer ($10^{-9}$ s) is much shorter than the time required for energy dissipation by relaxation processes ($10^{-5}$ s). Consequently, a non-equilibrium state develops in molecular size domains, possessing an instantaneous temperature that is much higher than the bulk temperature, and affects reaction process [19, 124].

Interpretations generally disagree, whether local thermal alterations may occur statistically in the bulk phase of the mixture [120] (due to fundamental inhomogeneities in the absorption process) or require the presence of well-absorbing reactants in the solution. The first possibility is more general, since it allows the heating of the solvent molecules as well (whose higher temperature will cause the reactant to overcome the activation energy), while the second one is limited to a few special systems with appropriately chosen polar reagents. The physical basis of the latter process was extensively discussed [123].

Dielectric relaxation spectroscopical studies on slow-moving (supercooled) liquids using low-frequency electric field provided evidence for existence of difference between effective temperature of a MW absorbing molecules and the bulk phase [121, 131]. Based on this observation, a model was constructed, which considers transient domains with configurational temperatures that are exceeding the measurable bulk temperature. This was adopted by Dudley for multicomponent solutions containing polar MW absorbing species [122, 123, 132, 133], and used to explain the results of MW assisted organic reactions.

A system consisting of dipolar molecules solvated in a non-polar, non-absorbing solvent is macroscopically homogeneous, but the solvent cages produce transient heterogeneity at the molecular level, which leads to spatially varying MW absorptivity (Fig. 4.1). Consequently, an absorbing domain forms in the vicinity of the polar molecule, within which heat is generated by MW dielectric loss processes. This local heating model is related to the concept of molecular radiators [134], which is defined as "a strongly absorbing solute that converts applied MW energy into thermal energy, generating heat in an otherwise poorly MW-absorbing system" [122].

In steady-state condition, the MW absorbing domain (polar molecules and their solvent cage) receive heat ($Q^{MW}$) through interaction with the electromagnetic field. The inbound heat is dissipated by conductive heat flow in the direction of the bulk medium ($Q^{med}$) and then out of the system to the surroundings ($Q^{sur}$). In this process, a temperature gradient necessarily forms (Fig. 4.1), in which the absorbing domain has higher effective temperature ($T^{dom}$) than the rest of the reaction medium ($T^{med}$). Naturally, both temperatures are higher than the vessel's environment ($T^{sur}$). It is important to emphasize that the system is not at thermal equilibrium, thus solution temperature is not indicative of the solute's kinetic energy.

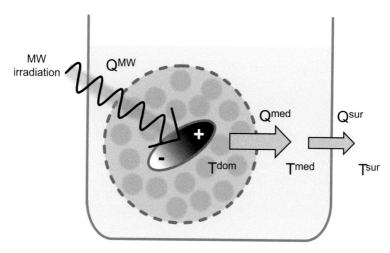

**Fig. 4.1** Energy transfer to MW absorbing domains in a non-absorbing medium

It was emphasized that this model is only applicable to specific systems, in which internal heat transfer is relatively inefficient, and the difference in MW absorptivity is large enough to allow accumulation of heat in the absorbing domain. Moreover, in completely insulated systems, the lack of outward heat transfer ($Q^{sur}$) prevents reaching the described steady-state.

Specifically designed "MW-actuated" reaction systems were used to test this hypothesis. Such homogeneous transformations were proposed, in which the chemical reactivity of highly MW-absorbing solutes could be investigated in an otherwise non-absorbing solvent [132]. The Friedel-Crafts type benzylation

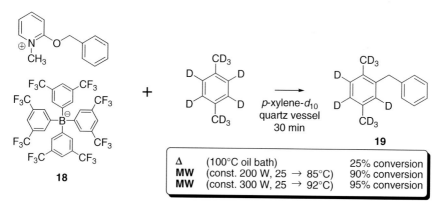

**Scheme 4.11** Benzylation of *p*-xylene

reaction (Scheme 4.11) utilizes[6] the ionic reagent (18), which interacts strongly with the MW field. On the other hand, the MW transparent $p$-xylene-$d_{10}$ acts as solvent[7] and reaction partner. After unimolecular thermolysis, the resulting benzyl cation leads to a single product (19), when reacting with $p$-xylene [132].

The first experiments were conducted in a temperature controlled MW setup. In the initial period—while high MW power was applied—rapid progress of the reaction was evidenced. However, after the control mechanism decreased power, the reaction rate diminished. This led the authors to recognize the importance of MW power level and utilization of irradiation with constant MW power (steady-state MW irradiation). Applying constant 200 W or 300 W MW power, significantly higher conversions were reached compared to the conventional control experiments. It is noteworthy that the final temperatures in MW heated cases were found to be lower, than in the oil bath heated counterpart.

However, the same reaction failed to show the same behavior under MW conditions at the hand of independent re-investigators [135]. In the subsequent debate [72, 73], the importance of the experimental fine details was elaborated. Later, a vast amount of experimental data was published by the original authors [122] to reinforce the findings, and kinetic data was used to estimate the magnitude of the extra-thermal effect. The effective temperature (average kinetic molecular energy) of the solute was found to be 20 °C higher, when exposed to constant MW irradiation.

In a related system [133], allyl $p$-nitrophenyl ether (20) was chosen as starting material, which is an efficient MW absorber. The first order reaction rate leading to the rearrangement product (21) showed minimal acceleration, when constant temperature MW conditions were used, but considerable increase was found in case of constant MW power irradiation (Scheme 4.12).

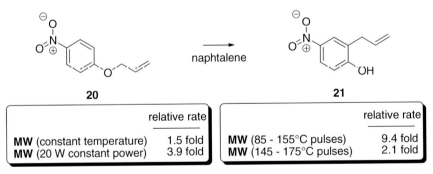

| | relative rate | | relative rate |
|---|---|---|---|
| **MW** (constant temperature) | 1.5 fold | **MW** (85 - 155°C pulses) | 9.4 fold |
| **MW** (20 W constant power) | 3.9 fold | **MW** (145 - 175°C pulses) | 2.1 fold |

**Scheme 4.12** Effect of pulsed MW irradiation on a Claisen rearrangement (relative rate is based on the comparable thermal treatment)

---

[6]The bulky anion was chosen for solubility reasons.

[7]Deuterated solvent was used for convenient NMR conversion determination.

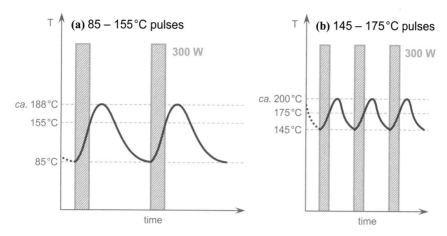

**Fig. 4.2** Different MW pulse programs

Further increase by pulsed MW irradiation was evidenced, compared to the calculated effect of the same temperature program applied conventionally. Interestingly, pulses in a lower temperature interval (programmed 85–155 °C, overshoots to *ca.* 188 °C, average temperature 130 °C, Fig. 4.2a) showed more pronounced effect, than the higher temperature range (programmed 145–175 °C, overshoots to *ca.* 200 °C, average temperature 172 °C, Fig. 4.2b). This could be explained by the fact that in the latter case a larger portion of the reaction occurs by thermal activation in the periods between the MW pulses, due to the higher temperatures thereof.

Pulsed MW irradiation was also found beneficial in an other instance of Claisen rearrangement [136], and in the course of natural product synthesis [137, 138] as well.

Related to the concept of molecular radiators, the MW specific activation (Scheme 4.13) in a ring-closing metathesis (RCM) reaction [139] should also be noted. The following experimental [140] and computational [141] studies confirmed the selective heating of the diene substrate (**22**), but also concluded that it is not instrumental to the progress of the reaction, as high conversion of the cyclic product (**23**) could also be obtained by conventional heating under strictly controlled conditions.

**Scheme 4.13** Ring-closing metathesis reaction (Ts: tosyl group; Mes: mesityl group; Cy: cyclohexyl group)

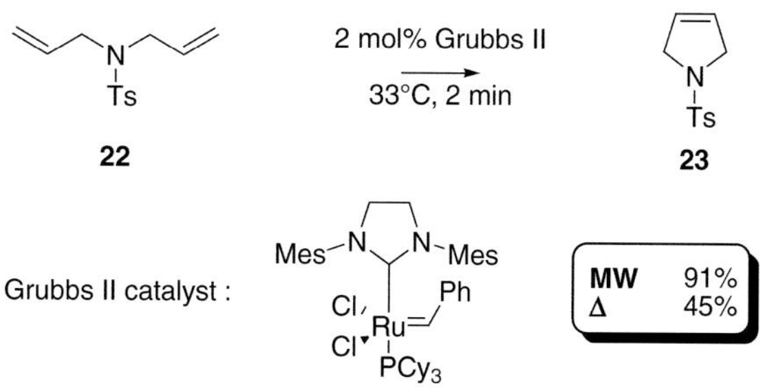

| | | |
|---|---|---|
| | **MW** | 91% |
| | Δ | 45% |

In conclusion, theory and experimental data suggest that localized thermal phenomena are not general, but particular reacting systems can in fact benefit from these microscopic level heating events, when suitable conditions are applied. Further research can help clearing confusion about the exact nature of these "hot-spot" related phenomena, provide direct proof on their existence, and broaden the scope of organic chemical reactions, in which these can be utilized for synthetic purposes.

## 4.6  Theories of Non-thermal MW Effects

Non-thermal (or athermal) MW effects describe changes in chemical transformations that cannot be explained by any forms of thermal differences [11–13, 142]. Based on the unique nature of MW heating, a few authors argued that such circumstances should exist, when temperature independent, direct interaction between the MW field and components of the reaction mixture plays a significant role. In other words, non-thermal MW effects and the physical changes leading to their development are independent from the well-known heating influence of the MW field. Presumably, the magnitude of these effects is dependent on MW field strength or irradiation power, although the nature of this relation hasn't been discussed.

The question of such effects has always been controversial, marked by non-fading disputes [72, 73, 132, 135, 143, 144], although a consensus [12, 20] seems to be reached on the non-existence of these effects. However, one should keep an eye on these theories as well, since these might inspire more correct rationalizations in the future. We don't aim to enumerate the plethora of claims (which have been reviewed extensively [9, 13, 14, 70, 142, 145, 146]) associated with non-thermal MW activation. Instead, we concentrate on showing the diversity of the explanations.

Despite clear theoretical distinctions, experimental separation of thermal and non-thermal effects is difficult. Reports of non-thermal effects are often contaminated by the simultaneously occurring thermal phenomena [36]. In order to study purely non-thermal effects, temperature has to be measured and controlled in the most precise manner both in dielectrically and conventionally heated parallel experiments [22, 63, 117, 118, 147, 148].

The special technique of MW heating with simultaneous cooling [124, 149] should be mentioned in relation to non-thermal MW effects. The sample is continuously cooled externally, while being irradiated using MWs, this way higher MW power levels can be achieved, while the same average temperature is maintained and overheating is avoided by removing excess heat. Since non-thermal effects are hypothesized to be dependent on the magnitude of the applied MW power or field strength, this methodology is thought to emphasize non-thermal MW effects at identical temperatures, and is a useful method for investigating thereof. However, when this technique is utilized, one should keep in mind that external IR thermometry is inadequate to correctly measure the temperature of the bulk reaction mixture, internal fiber-optic (FO) temperature probes should be used instead [20, 148, 150].

## 4.6.1    Changes in Kinetic Parameters

Studies aiming to establish the kinetic parameters in MW irradiated reacting systems often found that the parameters of the Arrhenius equation significantly differed from those, which were measured in the conventionally heated control experiments. Changes in both the pre-exponential factor ($A$) [145, 151–153] and the activation energy ($E_a$) [4, 126, 154] were demonstrated.

Improvements in the *pre-exponential term* require more frequent or more efficiently oriented collisions. Higher mobility of ionic species, leading to increased collision rate were suggested in MW assisted synthesis of inorganic materials [151, 155]. Speculations on orienting effects leading to a specific arrangement of molecules, in which the dipole vectors are more-or-less parallel to the applied electrical field gained notable attention. This way, polar reactants could be pre-organized in a way that results in more beneficial collisions, thus increasing the value of the pre-exponential term. However, theoretical considerations reveal [18, 156, 157] that thermal motion of the molecules overrides electrically induced orientation at the applied filed strengths.

Decrease in the *activation energy* was observed in several specific reactions. The apparent change however, could be rationalized by thermal alterations [126] in some cases. The alteration of the free energy of activation ($\Delta G^{\#}$) was hypothesized to originate from the change of the system's entropy [4]. Rapid rotation of dipoles was speculated to increase the probability molecular contact, thus lowering activation energy [154].

The higher reaction rates observed in the hydrolysis of sucrose were found to originate from beneficial changes in both activation energy and pre-exponential factor, explained by changes in ground state vibrational levels of the molecules [158].

The non-thermal influence of MW irradiation on the kinetics of transport processes was also considered. High efficiency in the solid-phase peptide synthesis was attributed to "molecular agitation" or "molecular stirring", which could replace mechanical stirring, due to the induced motion of molecules [159]. Enhancement of transport processes at interfaces in multiphase systems [92] has also been proposed, and used to explain rate accelerations in reactions where diffusion is the rate limiting step [145, 160].

## 4.6.2    Effects Connected to Polarity Change

A comprehensive explanation of the energetic changes was given in case of polar reaction mechanisms [70, 146]. It was observed that reactions with non-polar mechanism failed to show any improvement compared to conventional heating, while in case of polar mechanisms considerable acceleration was shown when the MW field was applied. It was concluded that the polarity change during the reaction

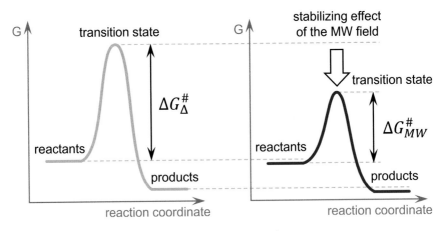

**Fig. 4.3** Theoretical effect of MW field on reaction energetics

governs the magnitude of MW effect, rather than the overall polarity of reacting species (which determines the heating characteristics). In specific reactions, significant development of charges occurs in the activated complex, compared to the ground state. The highly polarized transition states experience energetically beneficial interactions with the electric field (Fig. 4.3), leading to their stabilization and the decrease of activation barrier. The absolute energies of the ground states (i.e., reactants and products) are minimally influenced, consequently, reaction heat change remains unaffected.

This theory also explains [70] that in the case of reactions conducted in polar solvents (that are efficiently interacting with the electromagnetic field) MW effects are masked, while apolar solvents (that are practically transparent to MW irradiation) and solvent-free conditions [58] allow specific absorption of MW energy by the reactants.

Several demonstrative examples were enumerated in support of this approach [70, 146]. The practical observations were extended by quantum chemical calculations that showed good correlation with the experimental findings in different irreversible Diels-Alder type transformations [161]. However, independent re-investigations put some of these experimental results in question [63]. Additionally, an extensive survey of reaction mechanisms was conducted in the search for possible non-thermal effects related to polarity changes, but the employed quantum chemical approaches didn't provide evidence for the suggested influence of polarity, the role of thermal effects was emphasized instead [162].

Apart from polarity based considerations, analyzing the polarizability of transition states in dipolar cycloadditions of $C_{70}$ fullerene (**24**) and azomethine ylide (**25**) provided explanation for the product distribution under MW irradiation (Scheme 4.14). While conventional heating led to the mixture of three regioisomers (**26a-c**; substituted at $C_1$–$C_2$, $C_5$–$C_6$, $C_7$–$C_{21}$ positions, respectively), MW heating provided similar yield and gave the main products (**26a** and **26b**) in a slightly different ratio, but $C_7$–$C_{21}$ isomer (**26c**) was not formed [163].

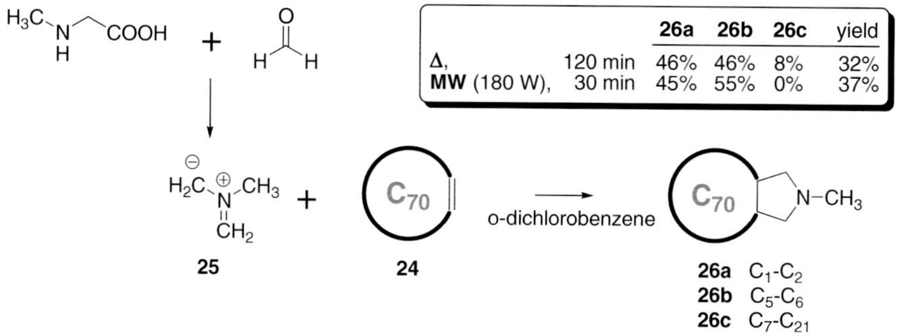

**Scheme 4.14** Cycloaddition reaction of $C_{70}$ fullerene

Based on computational studies, the addition of *N*-methylazomethineylide (**25**) was found to be stepwise, consisting of the nucleophilic addition of the 1,3-dipole resulting in zwitterionic intermediate species. The transition states and intermediates leading to the mentioned products were compared. The most polarized species ("hard", explicitly separated charges) were favored under MW irradiation. On the other hand, the thermally preferred product formed through stable, delocalized ("soft") transient structures. This theory could be used to explain the selectivities in further reactions involving polar transition states as well [164–166].

## 4.6.3   Interaction with MW Photons

Analogously to the quantum mechanical basics of photochemistry and molecular spectroscopy, the resonant type absorption of MW photons by specific molecules inspired some explanations to the non-thermal effects leading altered reactivity. In these hypothetical quantum resonance processes, the energy of the MW photon is thought to be directly transferred to the molecule, which is excited to a higher energy state as a consequence.

It is clear that unlike the UV photons, the energy associated with MW photons isn't sufficient to excite electronic states of the molecule or induce bond breakage [156, 157]. In fact, it is lower than the energy required to break hydrogen bonds, and also the average energy levels of thermal (Brownian) motion [18]. In a related sense, the excitation of vibrational or rotational state transitions (that are utilized in the conceptually unrelated MW spectroscopy), described as "spectroscopic heating" [167, 168] is also impossible, because the frequency of MW irradiation doesn't overlap with the resonant frequencies of molecular vibration and rotations [6].

Efficient coupling of specific bonds or functional groups with MW irradiation [169, 170], called as "antenna groups" [5] was also suggested. This way, directly transferring and storing excess energy in them, would lead to "activation" of individual bonds or functional groups and higher reaction rates, when these units of the molecule are participating in a reaction. These claims can be theoretically disproved, as energy is readily transferred to neighboring molecules trough relaxation processes [16].

Finally, these resonant processes are unlikely in condensed phases, because of the interplay of relaxation processes [123]. The MW heating increases the internal energy of the material which is partitioned among translational, vibrational and rotational energy regardless of the method of heating, instead of accumulating as rotational energy [39].

Regardless of the discussed fundamental considerations, the direct interaction of molecules and the electromagnetic field was studied using quantum mechanical methods. A possible role of MW influence on activation energy trough the rotational excitation on collision geometry [171] and its implications on hydrogen transfer kinetics [172] was theoretically analyzed. The catalytic consequence of rotationally excited states was studied in the computational simulation of the hydrolysis of an ester bond [173]. In the quantum chemical investigation of a specific $S_N2$ reaction, such vibrational motion (in the form of hindered rotation) was found, which overlaps with the normal MW heating frequency, suggesting that MW photons may excite these vibrational modes. It was emphasized that the solvation shell should be taken into account, as these molecular motions were negligible in gas-phase calculations [174].

In conclusion, the question of non-thermal effects has always been controversial, as neither direct evidence, nor provable theoretical explanation has been given on their existence. The general conclusion based on the available experimental data is that the non-thermal origin of MW phenomena should be generally rejected. However, it cannot be precluded that in some systems non-thermal MW effects can exist, but one should bear in mind that in order to demonstrate these kind of effects, only the most accurate comparison studies are suitable.

## 4.7  Summary

During the past three decades of MW assisted organic chemistry, the initial observations of unexpected reaction behavior obtained in MW reactors grew to a general understanding of MW effects. By exploiting the highly controlled conditions in modern MW equipment, it was realized that the majority of these phenomena can be explained by the effects of measurable temperature alterations. Macroscopic thermal effects could be demonstrated in homogeneous and heterogeneous systems, although advanced thermometry was required in the most challenging cases.

However, the outcome of a few exceptional MW assisted reactions couldn't be rationalized based on the measured temperature, which points to the possibility of microscopic thermal MW effects. The most fundamental characteristics and possible origin of the associated microscopic "hot-spots" were revealed by recent theoretical and experimental investigations. The concept of non-thermal MW effects was also suggested to explain these observations, although it was rejected in general, because of the absence of proper theoretical fundaments.

In conclusion, the exact influence of MW heating on organic chemical reactions still poses an exciting question, which requires the most precise and careful experimental techniques to properly be answered.

# References

1. Gedye R, Smith F, Westaway K, Ali H, Baldisera L, Laberge L, Rousell J (1986) The use of microwave ovens for rapid organic synthesis. Tetrahedron Lett 27:279–282. doi:10.1016/S0040-4039(00)83996-9
2. Giguere RJ, Bray TL, Duncan SM, Majetich G (1986) Application of commercial microwave ovens to organic synthesis. Tetrahedron Lett 27:4945–4948. doi:10.1016/S0040-4039(00)85103-5
3. Gedye RN, Smith FE, Westaway KC (1988) The rapid synthesis of organic compounds in microwave ovens. Can J Chem 66:17–26. doi:10.1139/v88-003
4. Berlan J, Giboreau P, Lefeuvre S, Marchand C (1991) Synthese organique sous champ microondes: premier exemple d'activation specifique en phase homogene. Tetrahedron Lett 32:2363–2366. doi:10.1016/S0040-4039(00)79924-2
5. Laurent R, Laporterie A, Dubac J, Berlan J, Lefeuvre S, Audhuy M (1992) Specific activation by microwaves: myth or reality? J Org Chem 57:7099–7102. doi:10.1021/jo00052a022
6. Raner KD, Strauss CR, Vyskoc F, Mokbel L (1993) A comparison of reaction kinetics observed under microwave irradiation and conventional heating. J Org Chem 58:950–953. doi:10.1021/jo00056a031
7. Gedye RN, Wei JB (1998) Rate enhancement of organic reactions by microwaves at atmospheric pressure. Can J Chem 76:525–532. doi:10.1139/v98-075
8. Berlan J (1995) Microwaves in chemistry: another way of heating reaction mixtures. Radiat Phys Chem 45:581–589. doi:10.1016/0969-806X(94)00072-R
9. Langa F, de la Cruz P, de la Hoz A, Díaz-Ortiz A, Díez-Barra E (1997) Microwave irradiation: more than just a method for accelerating reactions. Contemp Org Synth 4:373–386. doi:10.1039/co9970400373
10. Kranjc K, Kocevar M (2010) Microwave-assisted organic synthesis: general considerations and transformations of heterocyclic compounds. Curr Org Chem 14:1050–1074. doi:10.2174/138527210791130488
11. Kappe CO (2004) Controlled microwave heating in modern organic synthesis. Angew Chem Int Ed 43:6250–6284. doi:10.1002/anie.200400655
12. Kappe CO, Stadler A, Dallinger D (2012) Microwave theory. In: Microwaves in organic and medicinal chemistry, 2nd edn. Wiley-VCH, Weinheim, pp 9–39. doi:10.1002/9783527647828.ch2
13. de la Hoz A, Díaz-Ortiz A, Moreno A (2005) Microwaves in organic synthesis. Thermal and non-thermal microwave effects. Chem Soc Rev 34:164–178. doi:10.1039/b411438h
14. de la Hoz A, Díaz-Ortiz A, Moreno A (2007) Review on non-thermal effects of microwave irradiation in organic synthesis. J Microw Power Electromagn Energy 41:45–47
15. Mingos DMP, Baghurst DR (1991) Tilden Lecture. Applications of microwave dielectric heating effects to synthetic problems in chemistry. Chem Soc Rev 20:1–47. doi:10.1039/cs9912000001
16. Gabriel C, Gabriel S, Grant EH, Halstead BSJ, Mingos DMP (1998) Dielectric parameters relevant to microwave dielectric heating. Chem Soc Rev 27:213–224. doi:10.1039/a827213z

17. Bogdal D (2005) Interaction of microwaves with different materials. In: Microwave-assisted organic synthesis, vol 25. Tetrahedron Organic Chemistry Series. Elsevier, Oxford, UK, pp 1–11. doi:10.1016/S1460-1567(05)80014-5

18. Stuerga D (2012) Microwave-materials interactions and dielectric properties: from molecules and macromolecules to solids and colloidal suspensions. In: de la Hoz A, Loupy A (eds) Microwaves in organic synthesis, 3rd edn. Wiley-VCH, Weinheim, Germany, pp 1–56. doi:10.1002/9783527651313.ch1

19. Hayes BL (2002) Microwave synthesis: chemistry at the speed of light. CEM Publishing, Matthews, NC

20. Schmink JR, Leadbeater NE (2010) Microwave heating as a tool for sustainable chemistry. In: Leadbeater NE (ed) Microwave heating as a tool for sustainable chemistry. CRC Press, Boca Raton, FL, pp 1–24. doi:10.1201/9781439812709-2

21. Stadler A, Kappe CO (2001) High-Speed couplings and cleavages in microwave-heated, solid-phase reactions at high temperatures. Eur J Org Chem 2001:919–925. doi:10.1002/1099-0690(200103)2001:5<919::AID-EJOC919>3.0.CO;2-V

22. Obermayer D, Kappe CO (2010) On the importance of simultaneous infrared/fiber-optic temperature monitoring in the microwave-assisted synthesis of ionic liquids. Org Biomol Chem 8:114–121. doi:10.1039/b918407d

23. Kremsner JM, Kappe CO (2006) Silicon carbide passive heating elements in microwave-assisted organic synthesis. J Org Chem 71:4651–4658. doi:10.1021/jo060692v

24. Bogdał D, Prociak A (2007) Fundamentals of microwaves. In: Microwave-enhanced polymer chemistry and technology. Blackwell Publishing Ltd, Oxford, UK, pp 3–32. doi:10.1002/9780470390276.ch1

25. Kremsner JM, Kappe CO (2005) Microwave-assisted organic synthesis in near-critical water at 300 °C—a proof-of-concept study. Eur J Org Chem 2005:3672–3679. doi:10.1002/ejoc.200500324

26. Leadbeater NE, Torenius HM (2002) A study of the ionic liquid mediated microwave heating of organic solvents. J Org Chem 67:3145–3148. doi:10.1021/jo016297g

27. Van der Eycken E, Appukkuttan P, De Borggraeve W, Dehaen W, Dallinger D, Kappe CO (2002) High-speed microwave-promoted Hetero-Diels–Alder Reactions of 2(1H)-Pyrazinones in ionic liquid doped solvents. J Org Chem 67:7904–7907. doi:10.1021/jo0263216

28. Hoffmann J, Nüchter M, Ondruschka B, Wasserscheid P (2003) Ionic liquids and their heating behaviour during microwave irradiation—a state of the art report and challenge to assessment. Green Chem 5:296–299. doi:10.1039/b212533a

29. Leadbeater N, Torenius H, Tye H (2004) Microwave-promoted organic synthesis using ionic liquids: a mini review. Comb Chem High Throughput Screen 7:511–528. doi:10.2174/1386207043328562

30. Habermann J, Ponzi S, Ley SV (2005) Organic chemistry in ionic liquids using non-thermal energy-transfer processes. Mini Rev Org Chem 2:125–137. doi:10.2174/1570193053544454

31. Hohmann E, Keglevich G, Greiner I (2008) The effect of onium salt additives on the Diels-Alder reactions of a 1-Phenyl-1,2-dihydrophosphinine oxide under microwave conditions. Phosphorus Sulfur Silicon Relat Elem 182:2351–2357. doi:10.1080/10426500701441473

32. Besson T, Kappe CO (2012) Microwave Susceptors. In: de la Hoz A, Loupy A (eds) Microwaves in Organic Synthesis, 3rd edn. Wiley-VCH, Weinheim, Germany, pp 297–346. doi:10.1002/9783527651313.ch7

33. Garrigues B, Laurent R, Laporte C, Laporterie A, Dubac J (1996) Microwave-assisted carbonyl Diels-Alder and Carbonyl-Ene reactions supported on graphite. Liebigs Ann 1996:743–744. doi:10.1002/jlac.199619960516

34. Nüchter M, Müller U, Ondruschka B, Tied A, Lautenschläger W (2003) Microwave-assisted chemical reactions. Chem Eng Technol 26:1207–1216. doi:10.1002/ceat.200301836

35. Razzaq T, Kremsner JM, Kappe CO (2008) Investigating the existence of nonthermal/specific microwave effects using silicon carbide heating elements as power modulators. J Org Chem 73:6321–6329. doi:10.1021/jo8009402

36. Obermayer D, Gutmann B, Kappe CO (2009) Microwave chemistry in silicon carbide reaction vials: separating thermal from nonthermal effects. Angew Chem Int Ed 48:8321–8324. doi:10.1002/anie.200904185

37. Gutmann B, Obermayer D, Reichart B, Prekodravac B, Irfan M, Kremsner JM, Kappe CO (2010) Sintered silicon carbide: a new ceramic vessel material for microwave chemistry in single-mode reactors. Chem Eur J 16:12182–12194. doi:10.1002/chem.201001703

38. Kappe CO (2013) Unraveling the mysteries of microwave chemistry using silicon carbide reactor technology. Acc Chem Res 46:1579–1587. doi:10.1021/ar300318c

39. Strauss CR, Trainor RW (1995) Developments in microwave-assisted organic chemistry. Aust J Chem 48:1665–1692. doi:10.1071/CH9951665

40. Strauss CR (2009) On scale up of organic reactions in closed vessel microwave systems. Org Process Res Dev 13:915–923. doi:10.1021/op900194z

41. Strauss CR, Rooney DW (2010) Accounting for clean, fast and high yielding reactions under microwave conditions. Green Chem 12:1340–1344. doi:10.1039/c0gc00024h

42. Damm M, Glasnov TN, Kappe CO (2010) Translating high-temperature microwave chemistry to scalable continuous flow processes. Org Process Res Dev 14:215–224. doi:10.1021/op900297e

43. Chebanov VA, Saraev VE, Desenko SM, Chernenko VN, Shishkina SV, Shishkin OV, Kobzar KM, Kappe CO (2007) One-pot, multicomponent route to pyrazoloquinolizinones. Org Lett 9:1691–1694. doi:10.1021/ol0704111

44. Siskin M, Katritzky AR (1991) Reactivity of organic compounds in hot water: geochemical and technological implications. Science 254:231–237. doi:10.1126/science.254.5029.231

45. Krammer P, Mittelstädt S, Vogel H (1999) Investigating the synthesis potential in supercritical water. Chem Eng Technol 22:126–130. doi:10.1002/(SICI)1521-4125(199902)22:2<126::AID-CEAT126>3.0.CO;2-4

46. Dallinger D, Kappe CO (2007) Microwave-assisted synthesis in water as solvent. Chem Rev 107:2563–2591. doi:10.1021/cr0509410

47. Polshettiwar V, Varma RS (2008) Aqueous microwave chemistry: a clean and green synthetic tool for rapid drug discovery. Chem Soc Rev 37:1546–1557. doi:10.1039/b716534j

48. Geuens J, Kremsner JM, Nebel BA, Schober S, Dommisse RA, Mittelbach M, Tavernier S, Kappe CO, Maes BUW (2008) Microwave-assisted catalyst-free transesterification of triglycerides with 1-Butanol under supercritical conditions. Energy Fuels 22:643–645. doi:10.1021/ef700617q

49. Moseley JD, Lenden P, Thomson AD, Gilday JP (2007) The importance of agitation and fill volume in small scale scientific microwave reactors. Tetrahedron Lett 48:6084–6087. doi:10.1016/j.tetlet.2007.06.147

50. Mingos DMP (2005) Theoretical aspects of microwave dielectric heating. In: Tierney JP, Lidström P (eds) Microwave assisted organic synthesis. Blackwell Publishing Ltd., Oxford, UK, pp 1–22. doi:10.1002/9781444305548.ch1

51. de la Hoz A, Díaz-Ortiz A, Moreno A (2004) Selectivity in organic synthesis under microwave irradiation. Curr Org Chem 8:903–918. doi:10.2174/1385272043370429

52. Kappe CO (2008) Microwave dielectric heating in synthetic organic chemistry. Chem Soc Rev 37:1127–1139. doi:10.1039/b803001b

53. Díaz-Ortiz Á, de la Hoz A, Carrillo JR, Herrero MA (2012) Selectivity modifications under microwave irradiation. In: de la Hoz A, Loupy A (eds) Microwaves in organic synthesis, 3rd edn. Wiley-VCH, Weinheim, Germany, pp 209–244. doi:10.1002/9783527651313.ch5

54. Yeboah KA, Boyd JD, Kyeremateng KA, Shepherd CC, Ingersoll IM, Jackson DL, Holland AW (2014) Large accelerations from small thermal differences: case studies and conventional reproduction of microwave effects on palladium couplings. Reac Kinet Mech Cat 112:295–304. doi:10.1007/s11144-014-0733-z

55. Abramovitch RA, Abramovitch DA, Iyanar K, Tamareselvy K (1991) Application of microwave energy to organic synthesis: improved technology. Tetrahedron Lett 32:5251–5254. doi:10.1016/S0040-4039(00)92356-6
56. Stuerga D, Gonon K, Lallemant M (1993) Microwave heating as a new way to induce selectivity between competitive reactions. Application to isomeric ratio control in sulfonation of naphthalene. Tetrahedron 49:6229–6234. doi:10.1016/S0040-4020(01)87961-8
57. Almena I, Díaz-Ortiz A, Díez-Barra E, de la Hoz A, Loupy A (1996) Solvent-Free Benzylations of 2-Pyridone. Regiospecific N- or C-alkylation. Chem Lett 25:333–334. doi:10.1246/cl.1996.333
58. Loupy A, Petit A, Hamelin J, Texier-Boullet F, Jacquault P, Mathé D (1998) New solvent-free organic synthesis using focused microwaves. Synthesis 1998:1213–1234. doi:10.1055/s-1998-6083
59. Varma RS (1999) Solvent-free organic syntheses. Green Chem 1:43–55. doi:10.1039/a808223e
60. Gawande MB, Bonifácio VDB, Luque R, Branco PS, Varma RS (2014) Solvent-free and catalysts-free chemistry: a benign pathway to sustainability. ChemSusChem 7:24–44. doi:10.1002/cssc.201300485
61. Keglevich G, Dudás E (2007) Microwave-promoted efficient synthesis of 2-Phosphabicyclo[2.2.2]octadiene- and Octene-2-oxides under solvent-free conditions in Diels-Alder reaction. Synth Commun 37:3191–3199. doi:10.1080/00397910701547532
62. Schanche J-S (2003) Microwave synthesis solutions from personal chemistry. Molec Divers 7:293–300. doi:10.1023/B:MODI.0000006866.38392.f7
63. Herrero MA, Kremsner JM, Kappe CO (2008) Nonthermal microwave effects revisited: on the importance of internal temperature monitoring and agitation in microwave chemistry. J Org Chem 73:36–47. doi:10.1021/jo7022697
64. Sturm GSJ, Verweij MD, van Gerven T, Stankiewicz AI, Stefanidis GD (2012) On the effect of resonant microwave fields on temperature distribution in time and space. Int J Heat Mass Transfer 55:3800–3811. doi:10.1016/j.ijheatmasstransfer.2012.02.065
65. Larhed M, Hallberg A (1996) Microwave-promoted palladium-catalyzed coupling reactions. J Org Chem 61:9582–9584. doi:10.1021/jo9612990
66. Dallinger D, Irfan M, Suljanovic A, Kappe CO (2010) An investigation of wall effects in microwave-assisted ring-closing metathesis and cyclotrimerization reactions. J Org Chem 75:5278–5288. doi:10.1021/jo1011703
67. Bond G, Moyes RB, Pollington SP, Whan DA (1991) The superheating of liquids by microwave radiation. Chem Ind 1991:686–687
68. Baghurst DR, Mingos DMP (1992) Superheating effects associated with microwave dielectric heating. J Chem Soc, Chem Commun 1992:674–677. doi:10.1039/c39920000674
69. Saillard R, Poux M, Berlan J, Audhuy-Peaudecerf M (1995) Microwave heating of organic solvents: thermal effects and field modelling. Tetrahedron 51:4033–4042. doi:10.1016/0040-4020(95)00144-W
70. Perreux L, Loupy A (2001) A tentative rationalization of microwave effects in organic synthesis according to the reaction medium, and mechanistic considerations. Tetrahedron 57:9199–9223. doi:10.1016/S0040-4020(01)00905-X
71. Chemat F, Esveld E (2001) Microwave super-heated boiling of organic liquids: origin, effect and application. Chem Eng Technol 24:735–744. doi:10.1002/1521-4125(200107)24:7<735::AID-CEAT735>3.0.CO;2-H
72. Dudley GB, Stiegman AE, Rosana MR (2013) Correspondence on microwave effects in organic synthesis. Angew Chem Int Ed 52:7918–7923. doi:10.1002/anie.201301539
73. Kappe CO (2013) Reply to the correspondence on microwave effects in organic synthesis. Angew Chem Int Ed 52:7924–7928. doi:10.1002/anie.201304368
74. Klán P, Literák J, Relich S (2001) Molecular photochemical thermometers: investigation of microwave superheating effects by temperature dependent photochemical processes. J Photochem Photobiol A 143:49–57. doi:10.1016/S1010-6030(01)00481-6

75. Dressen MHCL, van de Kruijs BHP, Meuldijk J, Vekemans JAJM, Hulshof LA (2007) Vanishing microwave effects: influence of heterogeneity. Org Process Res Dev 11:865–869. doi:10.1021/op700080t

76. Dressen MHCL, van de Kruijs BHP, Meuldijk J, Vekemans JAJM, Hulshof LA (2011) Variable microwave effects in the synthesis of ureidopyrimidinones: the role of heterogeneity. Org Process Res Dev 15:140–147. doi:10.1021/op100202j

77. Conner WC, Tompsett GA (2008) How could and do microwaves influence chemistry at interfaces? J Phys Chem B 112:2110–2118. doi:10.1021/jp0775247

78. Tsukahara Y, Higashi A, Yamauchi T, Nakamura T, Yasuda M, Baba A, Wada Y (2010) In situ observation of nonequilibrium local heating as an origin of special effect of microwave on chemistry. J Phys Chem C 114:8965–8970. doi:10.1021/jp100509h

79. Kabb CP, Carmean RN, Sumerlin BS (2015) Probing the surface-localized hyperthermia of gold nanoparticles in a microwave field using polymeric thermometers. Chem Sci 6:5662–5669. doi:10.1039/C5SC01535A

80. Turner MD, Laurence RL, Conner WC, Yngvesson KS (2000) Microwave radiation's influence on sorption and competitive sorption in zeolites. AlChE J 46:758–768. doi:10.1002/aic.690460410

81. Blanco C, Auerbach SM (2002) Microwave-driven zeolite–guest systems show athermal effects from nonequilibrium molecular dynamics. J Am Chem Soc 124:6250–6251. doi:10.1021/ja017839e

82. Blanco C, Auerbach SM (2003) Nonequilibrium molecular dynamics of microwave-driven zeolite–guest systems: loading dependence of athermal effects. J Phys Chem B 107:2490–2499. doi:10.1021/jp026959l

83. Hájek M (2006) Microwave catalysis in organic synthesis. In: Loupy A (ed) Microwaves in organic synthesis, 2nd edn. Wiley-VCH, Weinheim, Germany, pp 615–652. doi:10.1002/9783527619559.ch13

84. Will H, Scholz P, Ondruschka B (2002) Heterogene Gasphasenkatalyse im Mikrowellenfeld. Chem Ing Tech 74:1057–1067. doi:10.1002/1522-2640(20020815)74:8<1057::AID-CITE1057>3.0.CO;2-3

85. Zhang X, Hayward DO, Mingos DMP (2003) Effects of microwave dielectric heating on heterogeneous catalysis. Catal Lett 88:33–38. doi:10.1023/A:1023530715368

86. Durka T, Van Gerven T, Stankiewicz A (2009) Microwaves in heterogeneous gas-phase catalysis: experimental and numerical approaches. Chem Eng Technol 32:1301–1312. doi:10.1002/ceat.200900207

87. Zhang X, Hayward DO, Mingos DMP (1999) Apparent equilibrium shifts and hot-spot formation for catalytic reactions induced by microwave dielectric heating. Chem Commun 1999:975–976. doi:10.1039/a901245a

88. Seyfried L, Garin F, Maire G, Thiebaut JM, Roussy G (1994) Microwave electromagnetic-field effects on reforming catalysts. 1. Higher Selectivity in 2-Methylpentane Isomerization on alumina-supported Pt catalysts. J Catal 148:281–287. doi:10.1006/jcat.1994.1209

89. Perry WL, Katz JD, Rees D, Paffet MT, Datye AK (1997) Kinetics of the microwave-heated CO oxidation reaction over alumina-supported Pd and Pt catalysts. J Catal 171:431–438. doi:10.1006/jcat.1997.1824

90. Vanier G (2007) Simple and efficient microwave-assisted hydrogenation reactions at moderate temperature and pressure. Synlett 2007:131–135. doi:10.1055/s-2006-958428

91. Heller E, Lautenschläger W, Holzgrabe U (2005) Microwave-enhanced hydrogenations at medium pressure using a newly constructed reactor. Tetrahedron Lett 46:1247–1249. doi:10.1016/j.tetlet.2005.01.002

92. Leskovsek S, Smidovnik A, Koloini T (1994) Kinetics of catalytic transfer hydrogenation of soybean oil in microwave and thermal field. J Org Chem 59:7433–7436. doi:10.1021/jo00103a041

93. Parvulescu AN, Van der Eycken E, Jacobs PA, De Vos DE (2008) Microwave-promoted racemization and dynamic kinetic resolution of chiral amines over Pd on alkaline earth supports and lipases. J Catal 255:206–212. doi:10.1016/j.jcat.2008.02.005

94. Arvela RK, Leadbeater NE (2005) Suzuki coupling of aryl chlorides with phenylboronic acid in water, using microwave heating with simultaneous cooling. Org Lett 7:2101–2104. doi:10.1021/ol0503384

95. Baxendale IR, Griffiths-Jones CM, Ley SV, Tranmer GK (2006) Microwave-assisted Suzuki coupling reactions with an encapsulated palladium catalyst for batch and continuous-flow transformations. Chem Eur J 12:4407–4416. doi:10.1002/chem.200501400

96. Irfan M, Fuchs M, Glasnov TN, Kappe CO (2009) Microwave-assisted cross-coupling and hydrogenation chemistry by using heterogeneous transition-metal catalysts: an evaluation of the role of selective catalyst heating. Chem Eur J 15:11608–11618. doi:10.1002/chem.200902044

97. Glasnov TN, Findenig S, Kappe CO (2009) Heterogeneous versus homogeneous palladium catalysts for ligandless Mizoroki-Heck reactions: a comparison of batch/microwave and continuous-flow processing. Chem Eur J 15:1001–1010. doi:10.1002/chem.200802200

98. Bogdal D, Lukasiewicz M, Pielichowski J, Miciak A, Bednarz S (2003) Microwave-assisted oxidation of alcohols using Magtrieve™. Tetrahedron 59:649–653. doi:10.1016/S0040-4020(02)01533-8

99. Lukasiewicz M, Bogdal D, Pielichowski J (2003) Microwave-assisted oxidation of side chain Arenes by Magtrieve™. Adv Synth Catal 345:1269–1272. doi:10.1002/adsc.200303131

100. Stuerga D, Gaillard P (1996) Microwave heating as a new way to induce localized enhancements of reaction rate. Non-isothermal and heterogeneous kinetics. Tetrahedron 52:5505–5510. doi:10.1016/0040-4020(96)00241-4

101. Baghurst DR, Mingos DMP (1988) Application of microwave heating techniques for the synthesis of solid state inorganic compounds. J Chem Soc Chem Commun 1988:829–830. doi:10.1039/c39880000829

102. Kniep R (1993) Fast solid-state chemistry: reactions under the influence of microwaves. Angew Chem Int Ed 32:1411–1412. doi:10.1002/anie.199314111

103. Strauss CR, Varma RS (2006) Microwaves in green and sustainable chemistry. In: Larhed M, Olofsson K (eds) Microwave methods in organic synthesis. Springer, Berlin, Germany, pp 199–231. doi:10.1007/128_060

104. Varma RS, Baig RBN (2012) Organic synthesis using microwaves and supported reagents. In: de la Hoz A, Loupy A (eds) Microwaves in organic synthesis, 3rd edn. Wiley-VCH, Weinheim, Germany, pp 427–486. doi:10.1002/9783527651313.ch10

105. Raner KD, Strauss CR, Trainor RW, Thorn JS (1995) A new microwave reactor for batchwise organic synthesis. J Org Chem 60:2456–2460. doi:10.1021/jo00113a028

106. Nilsson P, Larhed M, Hallberg A (2001) Highly regioselective, sequential, and multiple palladium-catalyzed arylations of vinyl ethers carrying a coordinating auxiliary: an example of a Heck triarylation process. J Am Chem Soc 123:8217–8225. doi:10.1021/ja011019k

107. Baghbanzadeh M, Carbone L, Cozzoli PD, Kappe CO (2011) Microwave-assisted synthesis of colloidal inorganic nanocrystals. Angew Chem Int Ed 50:11312–11359. doi:10.1002/anie.201101274

108. Washington AL, Strouse GF (2008) Microwave synthesis of CdSe and CdTe nanocrystals in nonabsorbing alkanes. J Am Chem Soc 130:8916–8922. doi:10.1021/ja711115r

109. Washington AL, Strouse GF (2009) Selective microwave absorption by trioctyl phosphine selenide: does it play a role in producing multiple sized quantum dots in a single reaction? Chem Mater 21:2770–2776. doi:10.1021/cm900305j

110. Young DD, Nichols J, Kelly RM, Deiters A (2008) Microwave activation of enzymatic catalysis. J Am Chem Soc 130:10048–10049. doi:10.1021/ja802404g

111. Copty A, Sakran F, Popov O, Ziblat R, Danieli T, Golosovsky M, Davidov D (2005) Probing of the microwave radiation effect on the green fluorescent protein luminescence in solution. Synth Met 155:422–425. doi:10.1016/j.synthmet.2005.09.028

112. van de Kruijs BHP, Dressen MHCL, Meuldijk J, Vekemans JAJM, Hulshof LA (2010) Microwave-induced electrostatic etching: generation of highly reactive magnesium for application in Grignard reagent formation. Org Biomol Chem 8:1688–1694. doi:10.1039/b926391h

113. Gutmann B, Schwan AM, Reichart B, Gspan C, Hofer F, Kappe CO (2011) Activation and deactivation of a chemical transformation by an electromagnetic field: evidence for specific microwave effects in the formation of Grignard reagents. Angew Chem Int Ed 50:7636–7640. doi:10.1002/anie.201100856

114. Jahngen EGE, Lentz RR, Pesheck PS, Sackett PH (1990) Hydrolysis of adenosine triphosphate by conventional or microwave heating. J Org Chem 55:3406–3409. doi:10.1021/jo00297a083

115. Nüchter M, Ondruschka B, Weiß D, Beckert R, Bonrath W, Gum A (2005) Contribution to the qualification of technical microwave systems and to the validation of microwave-assisted reactions and processes. Chem Eng Technol 28:871–881. doi:10.1002/ceat.200500136

116. Durka T, Stefanidis GD, Gerven TV, Stankiewicz A (2010) On the accuracy and reproducibility of fiber optic (FO) and infrared (IR) temperature measurements of solid materials in microwave applications. Meas Sci Technol 21:45108–45108. doi:10.1088/0957-0233/21/4/045108

117. Hayden S, Damm M, Kappe CO (2013) On the importance of accurate internal temperature measurements in the microwave dielectric heating of viscous systems and polymer synthesis. Macromol Chem Phys 214:423–434. doi:10.1002/macp.201200449

118. Kappe CO (2013) How to measure reaction temperature in microwave-heated transformations. Chem Soc Rev 42:4977–4990. doi:10.1039/c3cs00010a

119. Mason TJ, Lorimer JP (1989) Sonochemistry: theory, applications and uses of ultrasound in chemistry. Ellis Horwood Series in Physical Chemistry. Ellis Horwood, New York, NY

120. Keglevich G, Greiner I, Mucsi Z (2015) An interpretation of the rate enhancing effect of microwaves—modelling the distribution and effect of local overheating—a case study. Curr Org Chem 19:1436–1440. doi:10.2174/1385272819666150528004505

121. Huang W, Richert R (2008) The physics of heating by time-dependent fields: microwaves and water revisited. J Phys Chem B 112:9909–9913. doi:10.1021/jp8038187

122. Rosana MR, Hunt J, Ferrari A, Southworth TA, Tao Y, Stiegman AE, Dudley GB (2014) Microwave-specific acceleration of a Friedel-Crafts reaction: evidence for selective heating in homogeneous solution. J Org Chem 79:7437–7450. doi:10.1021/jo501153r

123. Dudley GB, Richert R, Stiegman AE (2015) On the existence of and mechanism for microwave-specific reaction rate enhancement. Chem Sci 6:2144–2152. doi:10.1039/C4SC03372H

124. Hayes BL (2004) Recent advances in microwave-assisted synthesis. Aldrichimica Acta 37:66–77

125. Schmink JR, Leadbeater NE (2009) Probing "microwave effects" using Raman spectroscopy. Org Biomol Chem 7:3842–3846. doi:10.1039/b910591c

126. Lewis DA, Summers JD, Ward TC, McGrath JE (1992) Accelerated imidization reactions using microwave radiation. J Polym Sci, Part A Polym Chem 30:1647–1653. doi:10.1002/pola.1992.080300817

127. Keglevich G, Kiss NZ, Jablonkai E, Bálint E, Mucsi Z (2015) The potential of microwave in organophosphorus syntheses. Phosphorus Sulfur Silicon Relat Elem 190:647–654. doi:10.1080/10426507.2014.989430

128. Keglevich G, Kiss NZ, Mucsi Z, Körtvélyesi T (2012) Insights into a surprising reaction: The microwave-assisted direct esterification of phosphinic acids. Org Biomol Chem 10:2011–2018. doi:10.1039/C2OB06972E

129. Keglevich G, Kiss NZ, Körtvélyesi T (2013) Microwave-assisted functionalization of phosphinic acids: amidations versus esterifications. Heteroat Chem 24:91–99. doi:10.1002/hc.21068

130. Mucsi Z, Kiss NZ, Keglevich G (2014) A quantum chemical study on the mechanism and energetics of the direct esterification, thioesterification and amidation of

1-hydroxy-3-methyl-3-phospholene 1-oxide. RSC Adv 4:11948–11954. doi:10.1039/C3RA47456A

131. Huang W, Richert R (2009) Dynamics of glass-forming liquids. XIII. Microwave heating in slow motion. J Chem Phys 130:194509–194522. doi:10.1063/1.3139519

132. Rosana MR, Tao Y, Stiegman AE, Dudley GB (2012) On the rational design of microwave-actuated organic reactions. Chem Sci 3:1240–1244. doi:10.1039/c2sc01003h

133. Chen P-K, Rosana MR, Dudley GB, Stiegman AE (2014) Parameters affecting the microwave-specific acceleration of a chemical reaction. J Org Chem 79:7425–7436. doi:10.1021/jo5011526

134. Kaiser NF, Bremberg U, Larhed M, Moberg C, Hallberg A (2000) Fast, convenient, and efficient molybdenum-catalyzed asymmetric allylic alkylation under noninert conditions: an example of microwave-promoted fast chemistry. Angew Chem Int Ed 39:3595–3598. doi:10.1002/1521-3773(20001016)39:20<3595::AID-ANIE3595>3.0.CO;2-S

135. Kappe CO, Pieber B, Dallinger D (2013) Microwave effects in organic synthesis: myth or reality? Angew Chem Int Ed 52:1088–1094. doi:10.1002/anie.201204103

136. Baxendale IR, Lee A-L, Ley SV (2002) A concise synthesis of carpanone using solid-supported reagents and scavengers. J Chem Soc Perkin Trans 1(2002):1850–1857. doi:10.1039/b203388g

137. Durand-Reville T, Gobbi LB, Gray BL, Ley SV, Scott JS (2002) Highly selective entry to the azadirachtin skeleton via a claisen rearrangement/radical cyclization sequence. Org Lett 4:3847–3850. doi:10.1021/ol0201557

138. Baxendale IR, Ley SV, Nessi M, Piutti C (2002) Total synthesis of the amaryllidaceae alkaloid (+)-plicamine using solid-supported reagents. Tetrahedron 58:6285–6304. doi:10.1016/S0040-4020(02)00628-2

139. Mayo KG, Nearhoof EH, Kiddle JJ (2002) Microwave-accelerated ruthenium-catalyzed olefin metathesis. Org Lett 4:1567–1570. doi:10.1021/ol025789s

140. Garbacia S, Desai B, Lavastre O, Kappe CO (2003) Microwave-assisted ring-closing metathesis revisited. On the question of the nonthermal microwave effect. J Org Chem 68:9136–9139. doi:10.1021/jo035135c

141. Rodríguez AM, Prieto P, de la Hoz A, Díaz-Ortiz A, García JI (2014) The issue of 'molecular radiators' in microwave-assisted reactions. Computational calculations on ring closing metathesis (RCM). Org Biomol Chem 12:2436–2445. doi:10.1039/c3ob42536c

142. de la Hoz A, Díaz-Ortiz Á, Gómez MV, Prieto P, Migallón AS (2012) Elucidation of microwave effects: methods, theories, and predictive models. In: de la Hoz A, Loupy A (eds) Microwaves in Organic Synthesis, 3rd edn. Wiley-VCH, Weinheim, Germany, pp 245–295. doi:10.1002/9783527651313.ch6

143. Kuhnert N (2002) Microwave-assisted reactions in organic synthesis—are there any nonthermal microwave effects? Angew Chem Int Ed 41:1863–1866. doi:10.1002/1521-3773(20020603)41:11<1863::AID-ANIE1863>3.0.CO;2-L

144. Strauss CR (2002) Microwave-assisted reactions in organic synthesis—are there any nonthermal microwave effects? Response to the highlight by N Kuhnert. Angew Chem Int Ed 41:3589–3590. doi:10.1002/1521-3773(20021004)41:19<3589::AID-ANIE3589>3.0.CO;2-Q

145. Jacob J, Chia LHL, Boey FYC (1995) Thermal and non-thermal interaction of microwave radiation with materials. J Mater Sci 30:5321–5327. doi:10.1007/BF00351541

146. Perreux L, Loupy A, Petit A (2012) Nonthermal effects of microwaves in organic synthesis. In: de la Hoz A, Loupy A (eds) Microwaves in organic synthesis, 3rd edn. Wiley-VCH, Weinheim, Germany, pp 127–207. doi:10.1002/9783527651313.ch4

147. Hostyn S, Maes BUW, Van Baelen G, Gulevskaya A, Meyers C, Smits K (2006) Synthesis of 7H-indolo[2,3-c]quinolines: study of the Pd-catalyzed intramolecular arylation of 3-(2-bromophenylamino)quinolines under microwave irradiation. Tetrahedron 62:4676–4684. doi:10.1016/j.tet.2005.12.062

148. Hosseini M, Stiasni N, Barbieri V, Kappe CO (2007) Microwave-assisted asymmetric organocatalysis. A probe for nonthermal microwave effects and the concept of simultaneous cooling. J Org Chem 72:1417–1424. doi:10.1021/jo0624187

149. Hayes BL, Collins MJ (2004) Reaction and temperature control for high power microwave-assisted chemistry techniques. World Patent WO 04002617, 8 Jan 2004

150. Leadbeater NE, Pillsbury SJ, Shanahan E, Williams VA (2005) An assessment of the technique of simultaneous cooling in conjunction with microwave heating for organic synthesis. Tetrahedron 61:3565–3585. doi:10.1016/j.tet.2005.01.105

151. Binner JGP, Hassine NA, Cross TE (1995) The possible role of the pre-exponential factor in explaining the increased reaction rates observed during the microwave synthesis of titanium carbide. J Mater Sci 30:5389–5393. doi:10.1007/BF00351548

152. Haque E, Khan NA, Park JH, Jhung SH (2010) Synthesis of a metal-organic framework material, iron terephthalate, by ultrasound, microwave, and conventional electric heating: a kinetic study. Chem Eur J 16:1046–1052. doi:10.1002/chem.200902382

153. Qi X, Watanabe M, Aida TM, Smith RL (2010) Fast transformation of glucose and Di-/Polysaccharides into 5-Hydroxymethylfurfural by microwave heating in an ionic liquid/catalyst system. ChemSusChem 3:1071–1077. doi:10.1002/cssc.201000124

154. Shibata C, Kashima T, Ohuchi K (1996) Nonthermal influence of microwave power on chemical reactions. Jpn J Appl Phys Part 1(35):316–319. doi:10.1143/JJAP.35.316

155. Rybakov KI, Semenov VE (1994) Possibility of plastic deformation of an ionic crystal due to the nonthermal influence of a high-frequency electric field. Phys Rev B 49:64–68. doi:10.1103/PhysRevB.49.64

156. Stuerga DAC, Gaillard P (1996) Microwave athermal effects in chemistry: a Myth's autopsy. Part I: historical background and fundamentals of wave-matter interaction. J Microw Power Electromagn Energy 31:87–100

157. Stuerga DAC, Gaillard P (1996) Microwave athermal effects in chemistry: a Myth's autopsy. Part II: orienting effects and thermodynamic consequences of electric field. J Microw Power Electromagn Energy 31:101–113

158. Adnadjevic BK, Jovanovic JD (2012) A comparative kinetics study on the isothermal heterogeneous acid-catalyzed hydrolysis of sucrose under conventional and microwave heating. J Mol Catal A Chem 356:70–77. doi:10.1016/j.molcata.2011.12.027

159. Chen S-T, Chiou S-H, Wang K-T (1991) Enhancement of chemical reactions by microwave irradiation. J Chin Chem Soc 38:85–91. doi:10.1002/jccs.199100015

160. Antonio C, Deam RT (2007) Can "microwave effects" be explained by enhanced diffusion? Phys Chem Chem Phys 9:2976–2982. doi:10.1039/b617358f

161. Loupy A, Maurel F, Sabatié-Gogová A (2004) Improvements in Diels-Alder cycloadditions with some acetylenic compounds under solvent-free microwave-assisted conditions: experimental results and theoretical approaches. Tetrahedron 60:1683–1691. doi:10.1016/j.tet.2003.11.042

162. de Cózar A, Millán MC, Cebrián C, Prieto P, Díaz-Ortiz A, de la Hoz A, Cossío FP (2010) Computational calculations in microwave-assisted organic synthesis (MAOS). Application to cycloaddition reactions. Org Biomol Chem 8:1000–1009. doi:10.1039/b922730j

163. Langa F, de la Cruz P, de la Hoz A, Espíldora E, Cossío FP, Lecea B (2000) Modification of regioselectivity in cycloadditions to C70 under microwave irradiation. J Org Chem 65:2499–2507. doi:10.1021/jo991710u

164. Bose AK, Banik BK, Manhas MS (1995) Stereocontrol of β-lactam formation using microwave irradiation. Tetrahedron Lett 36:213–216. doi:10.1016/0040-4039(94)02225-Z

165. Arrieta A, Lecea B, Cossío FP (1998) Origins of the stereodivergent outcome in the staudinger reaction between acyl chlorides and imines. J Org Chem 63:5869–5876. doi:10.1021/jo9804745

166. Díaz-Ortiz A, de la Hoz A, Herrero MA, Prieto P, Sánchez-Migallón A, Cossío FP, Arrieta A, Vivanco S, Foces-Foces C (2003) Enhancing stereochemical diversity by means of microwave irradiation in the absence of solvent: synthesis of highly substituted nitroproline

esters via 1,3-dipolar reactions. Molec Divers 7:175–180. doi:10.1023/B:MODI.
0000006799.01924.2e

167. Sun WC, Guy PM, Jahngen JH, Rossomando EF, Jahngen EGE (1988) Microwave-induced
hydrolysis of phospho anhydride bonds in nucleotide triphosphates. J Org Chem 53:4414–
4416. doi:10.1021/jo00253a047

168. Pagnotta M, Pooley CLF, Gurland B, Choi M (1993) Microwave activation of the
mutarotation of α-D-glucose: an example of an interinsic microwave effect. J Phys Org Chem
6:407–411. doi:10.1002/poc.610060705

169. Adámek F, Hájek M (1992) Microwave-assisted catalytic addition of halocompounds to
alkenes. Tetrahedron Lett 33:2039–2042. doi:10.1016/0040-4039(92)88135-R

170. Zijlstra S, De Groot TJ, Kok LP, Visser GM, Vaalburg W (1993) Behavior of reaction
mixtures under microwave conditions: use of sodium salts in microwave-induced N-[18F]
fluoroalkylations of aporphine and tetralin derivatives. J Org Chem 58:1643–1645. doi:10.
1021/jo00059a002

171. Miklavc A (2001) Strong Acceleration of Chemical Reactions Occurring Through the Effects
of Rotational Excitation on Collision Geometry. ChemPhysChem 2:552–555. doi:10.1002/
1439-7641(20010917)2:8/9<552::AID-CPHC552>3.0.CO;2-5

172. Miklavc A (2004) Kinetic isotope effect in hydrogen transfer arising from the effects of
rotational excitation and occurrence of hydrogen tunneling in molecular systems. J Chem
Phys 121:1171–1174. doi:10.1063/1.1774162

173. Bren U, Krzan A, Mavri J (2008) Microwave catalysis through rotationally hot reactive
species. J Phys Chem A 112:166–171. doi:10.1021/jp709766c

174. Kalhori S, Minaev B, Stone-Elander S, Elander N (2002) Quantum Chemical Model of an
SN2 Reaction in a Microwave Field. J Phys Chem A 106:8516–8524. doi:10.1021/
jp012643m

Printed in the United States
By Bookmasters